BEAST**TECH**

DEFENDER

CRANE, MO

BEAST TECH

Defender

Crane, MO 65633

©2013 by Thomas Horn

A collaborative work by Thomas Horn and Terry Cook.

All rights reserved. Published 2013.

Printed in the United States of America.

ISBN 13: 978-0-9848256-6-0

A CIP catalog record of this book is available from the Library of Congress.

Cover illustration and design by Daniel Wright.

All Scripture quotations from the King James Version.

Some material in this work is from Terry Cook's, *The Mark of the New World Order*, first published in 1995.

Contents

Foreword

By John McTernan

There is not a more spine-chilling section in the entire Bible than Revelation 13; this chapter describes a world dictator who appears just prior to the Second Coming of Jesus Christ. He is commonly known as the Antichrist or the Beast. This Beast uses a universal numbering system placed on people called "the Mark of the Beast." No one can buy or sell without this "Mark." It seems that everyone knows about the number "666," even those with little knowledge of the Bible, as it is identified as the infamous "devil's number." This is probably one of the most widely known verses in the entire Bible.

> And he causeth all…to receive a mark in their right hand, or in their foreheads: And that no man might buy or sell, save he that had the mark, or the name of the beast, or the number of his name. Here is wisdom. Let him that hath

understanding count the number of the beast: for it is the number of a man; and his number is Six hundred threescore and six [666]. (Revelation 13:16–18)

For centuries, the fulfillment of these verses was a mystery. No one could understand how a marking system would control buying and selling throughout the entire world. It no longer takes *simple faith* to believe this, as the literal fulfillment is now coming together with incredible speed right before your eyes.

Computers, the Internet, Wi-Fi, social media, and all modern technology, plus worldwide trends (such as a one-world economic system and government), are creating conditions that can fulfill Revelation 13:16–18. As the prophet Daniel wrote twenty-five hundred years ago, knowledge would explode at the very time the 666 surveillance system came together.

Today a person can use a credit card virtually any place in the world. It has become the universal financial instrument. This instrument is taken for granted, but think how technologically advanced transactions have become when a person can make a purchase in a foreign country without currency. This same purchase is then instantly registered and the funds withdrawn.

All merchandise now is identified with a RFID (radio-frequency identification) label or the QR (Quick Response) code for electronic tracking. A smart phone can read the QR code. The International Business Machines Corporation (commonly referred to as "IBM") is now working to identify and track every item manufactured and sold in the world; these numbers are in the trillions. Then it will be very easy to match these items with the purchaser. With the rapidly advancing technology, this is not far off.

The technological ability to operate a global, cashless economic system is now in place for a good percentage of the world. The elimination of cash is all that is needed to bring this system online. A worldwide economic crisis could be the catalyst to set this up.

The Bible states that no one is going to buy or sell without a "mark in the right hand or forehead." This Mark is in the advanced developmental stage with various applications to put "electronic tattoos" on the body. One idea is to merge technology with bio-technology. The microelectronics technology is called an epidermal electronic system (EES). The idea is to have a substance like silk-laced microelectronics that dissolves and leaves the circuits on the skin. This system is planned to be tied into a universal Wi-Fi. With this structure in full operation, there could be real-time monitoring of everything being sold. No one will be able buy or sell without government approval.

Tied into the ability to control all buying and selling will be the complete monitoring of everything related to you. The Mark of the Beast is a numbering system, and now, through your Social Security number (SSN), all information about you can be monitored. You need this number for everything from bank accounts to a driver's license, along with all medical information and credit cards. Everything now revolves around your SSN. With the added advantages of newly developing technology, everything controlled by the SSN will be centralized into one file and controlled and updated in real time. This is not theory; it is on the way.

Starting in 2013, the federal government has the Utah Data Center (UDC) nearly in full operation. This is a $2 billion complex with one hundred thousand square feet of computer space. All of the Global Information Grid will be routed through the UDC with computers that can store yottabytes (10^{24} bytes). Everything that is digitalized about you will be stored there, down to parking tickets. Nothing will be left out.

To handle this vast information, the computers will operate at petaflop speed a second (10^{15}; floating point operations per second). To transmit this vast amount of information, researchers invented chips with built-in lasers that use multiple wavelengths of light that can transmit data at terabit speed per second (10^{12} bits transmitted every second).

In addition to controlling buying and selling, we are now heading into a time of a total surveillance society. Every day, a vast number of surveillance cameras are being installed throughout America. Every mall is under total surveillance. By 2015, the government plans for thirty thousand surveillance drones in the sky. There are now x-ray machines that can peer through the walls of your home and auto. The government is using robot insects wired with microphones for spying. There is now software that allows tracking hundreds of cell phones at once and reads tens of thousands of emails. Homeland Security recently admitted to monitoring Facebook, Twitter, and Google. All the high tech in your vehicle and cell phone can easily be tracked and stored for your location. All this will be stored in the UDC.

The rebuilding of the World Trade Center (WTC) shows how advanced and all-encompassing the 666 surveillance system has become. Surveillance equipment will cover the entire area, with state-of-the-art technology used, including facial recognition systems, retina scanners, and fully-automated, highly "intelligent" cameras that use artificial intelligence along with software to automatically detect any "unusual" movements. All people movements will be tracked by the computers to detect suspicious behavior and alert the police when necessary. There will be infrared sensors and heat detectors (capable of spotting explosives) alongside radiation detectors. The WTC is a prototype of what is coming everywhere as I write. Picture this surveillance system citywide, statewide, and nationwide.

All this surveillance information can be instantly tied into the Internet through the coming national Wi-Fi. The ability for complete surveillance of your life is just a few technological advancements away. As technology advances, just imagine how your life will come under complete surveillance. The 666 surveillance society is going to be all inclusive, and this is leading to an incredible, high-tech mouse/man trap with no way out. This is exactly what the Bible said would happen: No one will buy or sell without government approval.

A world dictator, the Antichrist, will soon come to power and use the 666 surveillance system to require everyone to worship him. Whoever resists will be killed. The Mark placed on (or in) the body will be the initiation into this system, which includes worshiping the Antichrist. This means the total rejection of Jesus Christ. The Bible warns that everyone who takes this Mark of the Beast will be damned for eternity. Once one receives it, there is no hope; his or her fate is sealed forever. Please keep that in mind as you read the following pages. Right now, you can choose Jesus Christ as your Savior, and I urge you to do so. Tomorrow will be a different set of situations as the 666 surveillance system—as thoroughly documented in *Beast Tech*—is coming at petaflop speed.

Introduction

By Thomas Horn

In the days leading up to the reelection of President George W. Bush in 2004, in one of several conversations I had with a fellow evangelical who couldn't understand my waning enthusiasm for the administration, I listened as the highly educated and very successful publisher said the dumbest thing I'd heard in a while—a statement that is unfortunately being repeated by many sheeple these days.

"But Tom," he said in response to my concern about the erosion of civil liberties, "if you don't have anything to hide, you don't have anything to worry about!"

Maybe it was because he had recently returned from a private dinner with then Attorney General John Ashcroft—which included a night of bulletproof limousine rides and exquisite dining locations under the ever-present protection of secret service agents—that my friend seemed asleep at the cerebral wheel.

Or perhaps it was due to the three of us—me, my friend, and Ashcroft—having history and credentials in the same organization, or the fact that one of my coauthors was an old college pal of John's, or Ashcroft's winning personality and the afterglow of the piano playing and gospel-songs crooning they had shared at the attorney general's private residence that night that had disarmed my buddy's brain.

After all, I'm sure all of this would make it difficult for any hard-core evangelical to understand my point of view: that the nation's top cop was at that time a scary man working with a group of even scarier men on some pretty scary policies.

Whatever the case, in the days following the election, I was pleased to learn that Ashcroft was stepping down. Yet, my bubble burst when insiders told me the celebration had come a bit early, and that Ashcroft's pro-abortion replacement was bringing some equally menacing qualities of his own to the thoroughly neo-conned White House. Multilateralism was being deemed outmoded, they said, and the new assertion would be that freedom from terrorism would only occur through preemptive action against "enemies of democracy" not only abroad, but also inside the United States.

I knew what they meant, and that it was bound to come sooner or later. I even predicted it—not that I'm a prophet, you understand, and not that anybody was listening way back then.

It was 1990 when, as keynote speaker to a packed house of religious delegates (including TBN [Trinity Broadcasting Network] officials), I howled, "If America doesn't wake up and teach its kids to think, we will have a dictator in this country within ten years!"

I still remember the looks on their faces, as if they were musing: *Here he goes with that "loss of liberty" speech again. Does Tom really believe that in just ten years Americans will be microchipped, body searched at airports, and detained without warrants?! Doesn't he know his "evidence-gained-through-torture-will-soon-pierce-the-heart-of-the-Geneva-Convention" line is getting old?!*

Of course they were right. I miscalculated. A dictator didn't take control of the US within ten years like I'd predicted. It now appears he will need a bit more time. But with leading members of the increasingly political and spiritually anemic evangelical community now spouting phrases like, "If you don't have anything to hide, you don't have anything to worry about," it shouldn't take too long. The eradication of fundamental rights—including the presumption of innocence, which my friend either didn't understand or didn't cherish—will be easy fodder for the emerging Beast as he soon plows the inevitable intrusion by federal bureaucrats into every corner of our lives.

Students of the Revolutionary War and the French Revolution will see the irony here. In 1789, while Americans were at long last rejoicing in newfound religious liberties and hard-won freedoms, more than twenty thousand citizens prepared to be executed in Paris' guillotines beneath the horrific persecution and torture of Maximilien Robespierre. France would witness an unprecedented reign of terror leading up to totalitarianism and Napoleon (whose name, ironically, means "the new Apollo" or "spirit of Antichrist" in the New Testament), in which the masses would in effect be told, "If you don't have anything to hide, you don't have anything to worry about!"

In case you're interested in how today's preemptive action

against "enemies of democracy inside the United States" might likewise materialize, here are the signs I warned of—right from my sermon notes—twenty-three years ago:

Signpost #1—The erosion of privacy:

As you learn to march in cadence to the New American Dream, changes to federal privacy laws will see you watched, monitored, investigated, patted down, detained, harassed, and suspected of criminal activity until you can prove otherwise. To ensure everybody's uncontested compliance along the new chain-cobbled highway, national ID and exhaustive databases will be networked to track continuous individual whereabouts and activities. Your car, television, computer, sidewalk, and building cameras will assure that Big Brother's benevolent eyes are watching everything you do.

Signpost #2—The erosion of property rights:

During the coming years, personal property rights will be vastly undermined. Why? Because of sustainable society ideas and also because when government accumulates power in responding to a crisis, it brings with it the authority to commandeer resources. The roof over your head is yours only as long as the government doesn't want it; every Real Estate agent knows this. It's called eminent domain, and I watched my grandfather lose his campground as a result of this law. How long will it be before similar principles apply to those hard-earned bucks you have in the bank? Already there is a global push for biometric body-parts scanning in order to verify your rights to access your own money so that you can buy and sell. Does this sound prophetic to anybody but me?

Signpost #3—The erosion of fundamental fairness and US due process:

Election "mandates" that are given or assumed could waste little time changing valuable court resources. To "make this country safer and to protect our mutual interests," citizens could expect a gradual decline in constitutional rights, including the misuse of laws and regulations by government officials as well as the implementation of new rules for domestic engagement—such as detention without formal charges.

Signpost #4—The erosion of free speech:

Any criticism of the US president and members of the government are currently being viewed as hostile. To speak your mind or disagree in the current political climate can malign you as a terrorist or, at a minimum, an unpatriotic radical. Like those worried pilgrims who fled in their little boats so long ago to escape totalitarianism in search of America, in the near future, dissenters will be viewed as enemies of the state if they challenge the status quo.

Well, I was such a radical back then.

On the other hand, it does seem everything I worried about has come to pass, and has become much worse in recent days.

Just consider the headlines currently filling daily news stories in which patriotic American citizens are increasingly defined as enemies of the state if they dare challenge the status quo.

For instance, a recent article at the *Digital Journal* titled, "CTC Says Opposition to a New World Order is Terrorist Activity," states:

West Point is the U.S. Military Academy that trains, educates, and prepares Cadets for their service in the U.S. Army. The CTC (Combating Terrorism Center) at West Point… provides a unique terrorism-based education and since its creation, the program has received international recognition for its studies, reports, and teachings on terrorism and terrorist threats. A new report from the CTC, however, suggests that far right wing political activists, not radical Islamic groups, are the new terrorist threat in America and even goes so far as to say that those who oppose a "New World Order" are potentially violent terrorists. (Alex Allen, *Digital Journal*, February 2, 2013; viewable here: http://digitaljournal.com/article/342686)

Even more to the point was a recent article from the British newspaper, *The Guardian*, titled, "Chilling Legal Memo from Obama DOJ Justifies Assassination of US Citizens," in which Glenn Greenwald noted, "The most extremist power any political leader can assert is the power to target his own citizens for execution without any charges or due process, far from any battlefield. The Obama administration has not only asserted exactly that power in theory, but has exercised it in practice."

Greenwald goes on to cite several examples of US citizens being killed under Obama's orders, including a sixteen-year-old American citizen, and then adds:

Not only is the entire process carried out solely within the Executive branch—with no checks or oversight of any kind—but there is zero transparency and zero accountability. The president's underlings compile their proposed lists of who should be executed, and the president—at a charming weekly event dubbed by White House aides as "Terror Tuesday"—then chooses from "baseball cards" and decrees in total secrecy who should die. The power of accuser, prosecutor, judge, jury, and executioner are all consolidated in this one man, and those powers are exercised in the dark. (Glenn Greenwald, *The Guardian*, February 5, 2013; viewable here: http://www.theguardian.com/commentisfree/2013/feb/05/obama-kill-list-doj-memo)

Last but not least, as this book heads to editing, US Defense Department documents were obtained by Judicial Watch, a conservative watchdog group, that warns of "extremists." The government paper actually describes these as people who "talk of individual liberties, states' rights, and how to make the world a better place." The astonishing manuscript goes on in a section labeled "Extremist Ideologies" to say, "In U.S. history, there are many examples of extremist ideologies and movements. *The colonists who sought to free themselves from British rule and the Confederate states who sought to secede from the Northern states are just two examples*" (emphasis added). (See: "Defense Department Guide Calls Founding Fathers 'Extremist,'" *The Daily Caller*, August 23, 2013: http://dailycaller.com/2013/08/23/defense-department-guide-calls-founding-fathers-extremist/.)

Say what? The colonists that came to America to *escape* extremism and to demand personal freedoms, sovereign states (the United States of America), and constitutional protections from big government are now to be perceived as examples of radical terrorists? This seems to have been taken right from the warnings of my twenty-three-year-old sermon, which evangelical leaders at the time thought could never happen in this country. Yet, now it's at our doorstep.

Still, until I took time to investigate *Beast Tech* with my friend Terry Cook, even I had no idea how far down the rabbit hole we had fallen and how close we really are to a single event that could give rise overnight to the era of Antichrist and the Mark of the Beast.

It is my earnest prayer the research in this book awakens every reader to the need to draw near to Jesus Christ and to trust Him for their tomorrow.

An Overview

Some even believe we [the Rockefeller family] are part of a secret cabal working against the best interests of the United States, characterizing my family and me as "internationalists" and of conspiring with others around the world to build a more integrated global political and economic structure— one world, if you will. If that's the charge, I stand guilty, and I am proud of it.

—David Rockefeller, *Memoirs*, p. 405

Clichés…clichés…clichés. Some people speak in nothing else, but sometimes they seem to serve, better than anything else, to illustrate a point or make a difficult message more easily understood. I imagine throughout this book we will probably use our share, and perhaps more. For example, there's the old "love and marriage…you can't have one without the other" cliché. It makes the exact point we wish to convey when telling you about the purpose of this book. Just substitute the words "surveillance and control" for the words "love and marriage," and you'll get the appropriate image.

The world is in the process of a great shaking, and not just physically (although in fulfillment of prophecy we can expect much more than we presently are experiencing). We are being shaken together in order to "even out the playing field," as it were, both socially and economically. The only way to accomplish this is if each nation surrenders its individual sovereignty to the leadership of a world-governing body or leader, who will in turn see to it that all nations will capitulate, either voluntarily or by force. Naturally, all of this will be "for our own good and/or convenience." What will finally "shake out" will be that which we have heard so much in recent years: the New World Order.

Many nations would benefit greatly from this "evening out" of the playing field, which, of course, would be at the expense of many others. For them it will cost *everything*—economy, military, health, education, natural resources, self-rule…in other words, all personal liberty/freedom and possessions.

However, many are not deceived by the old line that they are "doing it for our own good," and are refusing to "roll over and play dead." Exactly what can you do about all these rabble-rousers and troublemakers? The only way for such a turnover of power to occur successfully is if everyone complies and cooperates. The "powers that be" have been planning this scenario for centuries. Now that the plan is about to see fruition, they are not going to look kindly upon dissenters or uncooperative types. (In a speech, Rockefeller stated that there was too little time left to get the plan accomplished gradually, because the fundamentalists were waking up, so they are stepping up their speed.)

The first thing the "powers" must accomplish is to remove as many as possible of our constitutionally protected rights and freedoms—legislatively and judicially—in effect, making all dissenters outlaws. Once they have succeeded in declaring their opposition to be the bad guys, then they have all the power of the police, military, and judiciary behind them. You will either join the system or oppose it. Of course, if you refuse to join, you are automatically in opposition and are thus one of the rabble-rousers or troublemakers mentioned above.

The only way that those who enforce the new regulations (for lack of a better word) will be able to achieve their goal of bringing everyone into the New World Order is through surveillance and control. They can't control you if they can't find you and keep track of your movements, activities, communications, purchases, et cetera. In other words, there must be a grand and global scheme for the *surveillance* of everyone before you can hope to *control* everyone.

An article by Richard Falk states, "It is evident that the new world order as conceived in Washington is about control and surveillance." Lest you be misled, do not mistake Mr. Falk for a right-wing conservative; on the contrary, he believes that their motives are anything but pure toward equality in the New World Order. In fact, he continues, the New World Order is "not about values or a better life for the peoples of the world" (Richard Falk, contributing author in: Tareq and Jacqueline Ismael, *The Gulf War and the New World Order: International Relations of the Middle East*, University

Press of Florida, 1994, 545). He believes the wealthy nations are using their technology as an advantage to keep the poorer nations from being a threat to them.

Although the major thrust of this book, *Beast Tech*, is identification and tracking systems, a certain amount of foundation needs to be laid first. Therefore, we will cover a number of subjects in addition to tracking technology.

Some people find this subject matter to be science fiction, some consider it fulfillment of Bible prophecy, and still others consider it the raving of some lunatic who sees a conspiracy under every rock; but the ones who study it carefully see it as the next logical step toward establishing the New World Order. Most, however, are somewhat staggered by the pervasiveness of the proposed systems and its seeming inevitability lurking just on the horizon. In other words, they are experiencing what has come to be known as "future shock." So we will touch on future shock and its effect on the masses, as well as on the individual.

We will address the New World Order: What is it? When is it? How will it take over? Who's in charge? Where did the idea originate, and how old is this plan for a One World Government?

Other subjects will include loss of privacy and electronic bondage, the United Nations Resolution 666, the new constitution, the Illuminati, biometrics, barcoding, GPS (Global Positioning System), the cashless society… Let's pause here for a minute. There are other subjects to mention, but we want to address this subject before we proceed, because we are far down the road toward a cashless society…farther than most of us realize.

This is one of the first few steps that must be enacted to get us ready for biochip implants, and it is an action that will be taken under the guise of what is "best for the people." If studied fairly, without consideration for the prophetic implications, there could possibly be a valid argument about whether a cashless society really is for our own good. However, the end result and drawbacks far outweigh any temporary convenience or good it will accomplish. Some of the positive theories and facts you will be hearing more and more about as the time draws closer to being a completely electronic-currency nation are that it will: inhibit terrorism; stop theft, robberies, and muggings; stop international drug dealing; stop income tax cheaters; prevent parents from ducking out of paying child support; and render obsolete the fear of misplacing a wallet or purse containing a roll of cash, among other things. Think of the convenience and safety.

But what are those drawbacks I mentioned? To start with, it probably does not come as a shock to you that everything you have, owe, or own is in a machine somewhere (including your health records, family history, religious affiliations, job history, credit history, political leanings, and much more), which gives anyone with the proper authorization all the information about your personal business. (Yet, currently, it does not take someone with privileged access or authorization to still tap into information about anyone else that should be considered "private" by the standards of yesteryear. A simple search of a non-famous individual's name on Google will easily prove my point. Have you looked up the satellite and street images of your home property on maps.google.com lately? If you haven't, it's more of an intimate view than you think, complete with the kids' pool toys and tree house, right where you left them, accessible by anyone with the Internet. And oh yeah, they didn't bother to blur out the license plate on your parked vehicle in the driveway, either.)

The identification system (SMART cards, MARC cards, or whatever) must be tied unalterably to you in order to prevent theft of the card and fraudulent use thereof. The card must be used in conjunction with some other source of positive identification, such as a retina scan or fingerprint match. Can you already see the direction this path is taking? What if your card were lost or stolen? We can fix that. This tiny little biochip no bigger than a grain of rice (and maybe smaller as technology progresses) can be slipped under your skin and give you lifetime access to everything the New World Order will afford its happy members.

Historically, the technology needed to unravel the Revelation 13:16–18 mystery of the Mark of the Beast in this light has been unavailable (no one can buy or sell without the Mark in his or her right hand or forehead). Naturally, the manufacturers of this technology try to assure us that it wasn't designed for that purpose, but even as they speak, plans are being discussed to implant the "difficult" cases, i.e., the prison population, runaway teens, the elderly who have a tendency to wander off and become lost, and even babies before they leave the hospital to put an end to kidnapping. Doesn't that sound like a *really GOOD thing*?

It is old news that the method is being used widely in pet identification and agricultural animals, but within recent years there were reports of tracking and punishing poachers of salmon with implants, and the government ruled that all surgically implanted body parts will henceforth contain a biochip (of course, that's also for your *good,* as it retains the information about its manufacturer, your surgeon, how and why it was implanted, etc.), and breast implants are some of the first to be manufactured with the biochip. Another use of radio transponder identification (biochips) was in permanently riveted vinyl bracelets applied to over fifty thousand Haitian and Cuban refugees at the Guantanamo Bay US military base. Since children's wrists were too small, they were applied to their ankles.

Have you noticed lately how everything on television is *smart?* Pay close attention to how the rash of commercials has played out over the last couple of decades—AT&T actually called its technology the "Smart Card," but the term "smart" is used in so many different advertisements that, if you start to pay attention to the evolution of that term in the media, it will amaze you. They want you to be so acquainted with the term (read that as "desensitized to the term") that it will seem like second nature to you to recognize when something is current.

Throughout this book, we will be discussing biochip technology—its origination, manufacture, current uses, proposed uses, and unlimited possible future uses. We will summarize these facts and introduce the prophetic connections as described in the Bible. We hope, by reading this book, you will be informed and awakened from your slumber—if, indeed, you have been blissfully, but ignorantly, slumbering along, enjoying all this new *Star Trek* technology without being cognizant of the dire consequences.

chapter one

Future Shock Is Here

Your wallet and credit cards are obsolete. You may as well leave them at home on a shelf near your dusty antique collection, reminding you of ancient history. While shopping for your shampoo and toothpaste, you need naught but to *wave your right hand* over a scanner at the checkout counter to pay for and collect your goods, as the amount due will be automatically withdrawn from your bank account.

But, of course, this is either too fantastical to be true, or too far off in the future for *you* to worry about. Right?

Survey says: Incorrect.

The technology we have established today already covers, and exceeds, such a feat. Take, for example, a standard laser scanner (be it handheld or fixed) at *any* store checkout. The reading of barcodes and Universal Product Codes (UPCs) by machine to report your purchases to a computer is already outdated news. The only old-fashioned cash registers in use in modern times are made of rainbow-colored plastic and produced by Fisher Price. How endearing.

"But just suggest something like an implant in human beings and the social outcry is tremendous," said ten years ago by Tim Willard, then executive officer of the World Future Society, a Washington, DC-based organization that claims twenty-five thousand members worldwide, including *Future Shock* author Alvin Toffler. "While people over the years have grown accustomed to artificial body parts, there is definitely a strong aversion to things being implanted. It's the *'BIG BROTHER is watching'* syndrome. People would be afraid that all of their thoughts and movements were being monitored. It wouldn't matter if the technology were there or not. People would still worry."

But wait, we're getting a bit ahead of ourselves here. First we need to define the term "future shock" and how it came into our vocabulary. According to *Merriam Webster's Collegiate Dictionary, Eleventh Edition*, "future shock" is defined as: "The physical and psychological distress suffered by one who is unable to cope with the rapidity of social and technological changes." As words in the dictionary go, this one is of fairly recent origin. The dictionary tells us that it first came into common usage about 1965.

Toffler cannot directly be credited with originating the term, but using it as the title of his book certainly promulgated its wide acceptance, and the definition is one with which most people of this decade are familiar, to at least some degree.

Future shock has come to be considered a negative term that

bridges the age gap from children (who feel they are being propelled into the future before they are ready, and that adults have pretty much messed it up for them anyway) to senior citizens (who are befuddled by all this electronic computer technology and long for the "good old days"). It also bridges all areas of life, from banking, economics, and industry to education, social sciences, art, music, communications, databases, etc. We are being hurled ever forward by knowledge that is expanding exponentially.

And the most distressing symptom of future shock is the dehumanizing feeling we experience that we are all becoming nothing more than numbers…and in fact, that is the truth. With the world shrinking because of travel and communication technology and the pressure by the "powers that be" to merge us all into a One World Government (spell that, New World Order), the only way to "organize" us and keep track of us (for our own good, of course), is to give us all a single identification method that can be recognized worldwide.

So, future shock is real…it's just not very much in the *future* anymore. Future shock is here…future shock is now. Perhaps "Big Brother" wasn't watching us by the year 1984, as Orwell's book title speculates, but he wasn't all that far off target. The technology is here now via fiber optics, satellites, and other modern marvels for Big Brother to sit on the information superhighway, right in your living room. The only difference is that during the time from 1984 to now, we have been brainwashed into believing that this is a wonderful thing, so Big Brother didn't have to force his way in, as Orwell postulated; he is with us by our own invitation.

Biochip—RFID Technology

The very idea of implanting *even our pets* with identifying micro-chip transponders was a concept that for years was approached very slowly. "We wanted to make sure it's right for the animals, and that the *community is willing to accept* this new technology," said Diane Allevato, director of the Novato, California, animal shelter. Now, however, such high-tech "tagging" is on "fast forward" since people have grown accustomed—by successful *conditioning*—to what once was considered offensive technology. For example, if people are now willing to have such devices injected into their pets, why would they not put them into their own bodies? Or, to soften the blow just a little, into the elderly or children to prevent wandering from home or kidnapping, prison inmates, and runaway teens, because all that would be for the good of society as a whole. Once you have edged your way into accepting these logical implant choices, it's no great leap to accept the fact that a nonremovable implant would be terribly convenient and much safer for us than carrying around cards for all our needs (economic, health, passports, travel, etc.)—cards that could be lost or stolen.

Biochip the Size of a Grain of Rice

For pets, the implanted chip is the size of a single grain of rice. It stores a number that identifies the pet's individual information (including the name of the pet, as well as the name, address, and phone number of the pet owner), which can be accessed by computer via a scan over the outside of the skin.

The Novato, California, animal shelter mentioned earlier was the first shelter in the country to use the biochips. One lady phoned the shelter to say she felt the implants were "unnatural and weird."

"And there is no doubt about it, injecting an animal with such a chip is a pretty unnatural thing to do," Allevato said. "But it's also unnatural, obscene really, that about 15 million stray animals are destroyed in this country each year because their owners cannot be found."

As of the writing of this manuscript, animal implants are no longer considered "unnatural and weird." In fact, the system is in common usage worldwide, and the "service" is readily available to pet owners at many veterinarians and animal hospitals. In addition, animal chipping has evolved from being a mere suggestion to becoming a mandate in most urban areas when pets are adopted from a shelter or bought from a store.

The transponder chips being implanted into most animals today are manufactured by four major companies: American Veterinary ID (AVID) of Norco, California; Infopet Corporation of Minnesota; Hughes Aircraft in Southern California; and Texas Instruments of Attleboro, Massachusetts. Originally, Destron IDI Corporation of Boulder, Colorado, began the concept, but it now serves primarily as a research and development company for both Hughes and Texas Instruments, through various joint ventures.

RFID Implants: Are We Ready?

Many more companies have now entered the RFID manufacturing field, and biochip usage now has surpassed the "personal pet" stage and is being used in many different areas of farming, ranching (no need for old-fashioned branding irons), racing, and wildlife identification and monitoring. In conjunction with computer technology, RFID biochips are used to track and monitor the health histories, bloodlines, etc., of various farm animals, race horses, breeding stock, *prize-breeding* stock, and more.

In the modern dairy business, good ol' Bessie can come into the barn, pass under a scanner, be directed to her stall, and find her own specially prepared diet in place, just awaiting her arrival—all based upon her identification combined with the computer-stored information about what the handlers want her to eat. Her health records, milk production, and other pertinent factors are all considered, and the optimum diet is assigned to her.

Wildlife has been tagged and tracked manually/visually for many years, but now it is possible to track life as small as a honeybee by attaching an RFID to its back. Of course, this is a real stroke of genius, according to the ecologists, for monitoring all types of wildlife, especially the endangered species.

According to Destron director Jim Seiler, implantable RFID biochips are even being used to track fish. In some fishery applications, salmon are injected with these biochips, then scanned and tracked as they pass through specially equipped dam sites "to assure environmentalists that they are not being chewed up in the dam's power turbines," Seiler said.

Following that, reports were coming out of Europe that poachers were being caught and fined because the biochip implants in the fish led the authorities to the poacher. The implications of this announcement were staggering. It meant that the biochip no longer had to be scanned at a close distance of three to twelve inches. It would indicate that technology has been developed and actually put into service that could scan for detection of biochips at great distances, possibly even by satellite. This was, in fact, the ultimate direction biochip technology was heading at the time, as surveillance and control of humans is of far more importance to the leaders of the New World Order than the tracking of animals. Biochips being manufactured now can send and receive data about location and personal information from anywhere in the world, and this technology not only exists in biochips, but a similar technology exists within the digital devices in every purse and back pocket in the country.

Take, for example, your smart phone or laptop (or any other portable convenience like electronic pads, notebooks, etc.). If your phone was purchased within the last couple of years from any major carrier, it likely came with a preinstalled GPS that keeps constant tabs on your travels. And, of course, that's handy when you need it for navigational purposes. What you may or may not know about this feature is that unless you go out of your way to turn the GPS in your phone *off* (it stays on perpetually otherwise), it attaches personal information to all other data-share occurrences on your phone.

Remember that picture you took of Aunt Suzie at the barbecue last weekend? Remember when you uploaded that picture to

Facebook and tagged all your friends in it? You left your GPS on in your phone at the moment the picture was taken, and now you, Aunt Suzie, and all others tagged in the photograph are shown as "present" at that barbecue event last weekend. The address and map directions to that exact location are provided at no extra cost by the click of a mouse (again, available to anyone with Internet access, *not* just "Big Brother"). Not only did you unknowingly report the address where *you* were last weekend, but you reported the location of Aunt Suzie and your friends.

If you chose to leave your personal home address where your spouse and children live omitted from the personal information tab on your Facebook account to ensure privacy, you just negated that safety step by taking a picture of Billy in his highchair and posting it online, forgetting of course to disconnect your phone's camera from the ever-present GPS. Now Billy can be tracked to his highchair, map and directions to your home included, by anyone online—and it does not stop simply with your phone's camera. *Whereas nobody may be referring to this cellular phone feature as a biochip implant, it's essentially a chip you voluntarily carry with you everywhere, reporting everything that you do and everywhere you've been.* If or when a mandatory, government-issued or world-government-issued implantable chip *does* become the new deal, the leap from your smart phone to the new chip in your hand will not be a long one.

While there are at least "10,000 application ideas to explore when it comes to the chip's potential," Seiler insisted, "Destron is only concerned with animal identification" and "*is not considering HUMAN application*" (emphasis added).

"There is no need to apply the technology to humans because

the human fingerprint is unique enough to identify them," Seiler said. "Animals don't have such a unique identifier." All present manufacturers deny that they are considering any form of human application; however, the latest info indicates to the contrary.

But Tim Willard, managing editor of the World Future Society's bimonthly magazine, *Futurists*, said the technology behind such a *human microchip* is "fairly uncomplicated and with a little refinement, *could be used in a variety of human applications.*... Conceivably, *a number could be assigned at birth* and go with a person throughout life" (emphasis added), Willard said.

"Most likely," he added, "it would be *implanted in the back of the right* or left *hand* for convenience, so that it would be *easy to scan* at stores.... It could be used as a *universal identification card* that would replace credit cards, passports, that sort of thing.... At the supermarket checkout stand, you would simply *pass your hand over the scanner* and your bank account would be automatically debited.... It could be programmed to replace a medical alert bracelet. For example, at the scene of an accident, a medic could *scan the victim's hand* to find out his recent medical history, allergies, a relative to contact, etc. This would be especially valuable if the person were unconscious" (emphasis added).

In another very logical application, such microchips could replace the need for keys to home, car, and workplace locks, since *the chip in one's hand* would serve as a *universal key* for all such locks. One would simply scan the back of his or her right hand over a high-tech microreader device designed into future locks to gain access. Security would be enhanced tremendously, since, allegedly, no one would be able to "pick" the locks, make a wax copy, or obtain a spare key found under a doormat. Doesn't that make you feel really safe? Not if you understand that any burglar worth his salt has always found a way around locks and security devices, no matter how complex. It might serve as a deterrent to an inept thief, but a serious crook would manage to get in somehow. The bigger the target, the cagier the crooks. Just think of any major malicious virus caused by such minds in the past ten years online…most designed by brilliant computer-degree-bearing hackers no more intimidated by a futuristic door lock than someone's email account.

New Biochips of Living Proteins and Hybrid Brain-Machine Interfacing

Some of the simple biochips of yesterday are now new biochips partially made of or with living proteins. According to Willard, "A powerful biochip…once surgically implanted in the brain, could make it possible to program or upload an unlimited amount of information into the mind, without ever having cracked a book…it will be infinitely smaller and have the capacity to carry much more information. It will have a wide range of functions that will simply boggle our minds."

After hearing more and more news of hybrid brain-machine interfaces (HMBIs), mind-machine interfaces (MMIs), and brain-computer interfaces (BCIs), Willard's comment seems almost tame. As far back as the 1970s, there has been a significant amount of focus put upon neuroprosthetics (artificial sensory intervention via prosthetics placed directly in the brain, essentially rerouting and/

or repairing the brain's communication with a body part due to birth defects, injury, or disease). After our country has observed the repeated miracles of such evolution between man and machine (literally observing the lame man walk, the blind man see, and the deaf man hear, among hundreds of other marvels of science), the fusion of a biochip with living proteins taking its podium in current labs is not hard to believe.

Loss of Freedom and Privacy

Chapter two of this book will discuss extensively the loss of our privacy and electronic bondage, but we should at least address it briefly here, as it contributes greatly to the *future shock* condition.

The capabilities described above carry with them the inherent risk of abuse by government and other organizations, particularly over the issue of privacy. How will access to such information be limited and controlled? And if (read that "when") it is passed around (often without our knowledge, much less our consent) to a variety of entities, both private and governmental, how will we know of errors that occur when someone is updating information? No system is foolproof...errors will *always* occur somehow. The one at fault for the error will not make any adjustments to *their* records/databases. You must determine the origin of the error and try your best to get them to correct the mistake (not likely to occur, we might warn you, since admitting such a blunder would make them financially liable for any injury you might have suffered because of their negligence). And getting them to pass such a correction on to everyone who received the erroneous data from them is practically hopeless.

It would be very difficult to impose enforceable limitations on the availability of such information in today's sophisticated environment of internationally linked databases. There would be no guarantees that confidential information might not inadvertently "leak" out to someone not authorized to have it...except YOU, of course. You are not authorized.

A Human Biochip ID System

A human microchip identification system, Willard said, "would work best with a highly centralized computer system where [only!] one identification number would gain access to medical, credit, academic, home security," and other kinds of data. But under this arrangement, as you can imagine, *security risks are somewhat intense.*

"People tend to be idealistic about their independence and privacy, but the reality is that most information pertaining to education, credit history, whatever, is readily available to just about anyone who asks. Anyone who has ever experienced a simple credit check knows this," Willard said.

One futurist found the concept of microchip implantation in humans offensive. "It reminds me of tattooing concentration camp victims during World War II," said Robert Mittman of the Institute for the Future, a nonprofit research and consulting firm in California. He said there were better methods of identifying people than "violating the integrity of their skin." He continued, "Personally, I have problems with it.… People would end up sacrificing some civil rights."

Another concerned individual is the associate director of the American Civil Liberties Union (ACLU) for Northern California, Martha Kegal. She expressed concern over how private records would be kept from "inquiring minds" if such a system existed.

The System Is Already in Place

But the question is not whether or not such a system exists, because we all know it does. It is already in place and obvious to all but the most naïve. Rather, the question most troubling any intelligent mind would seem to be the system's final destination, as it were. Whenever we get to wherever it ultimately is taking us…will it have proven to be good or evil? For an answer to this question, we must turn to our Creator's operations manual, the Bible, for only God knows how all things will end.

Only the Bible Has the Answers

According to God's Holy Word, the result of all our present circumstances is quite clear. Our generation may well be the one that has been appointed to be alive at the time of the conclusion of all ages, better known as the *end of the world*. For the first time in history, the technology is available to accomplish the things prophesied in the

Bible. Precisely which technology will be used is immaterial—the fact is, never prior to this generation has *any* technology been available to permit the conduct of commerce in the manner described in Revelation 13:16–17.

Such an end-times scenario necessarily involves the implantable biochip technology (or something comparable) around which this book has been written, for without it, the final worldwide system of satanic government, as outlined in the books of Daniel and Revelation, simply would not be feasible.

Current technological, social, and/or political circumstances soon will lead to a system of universal totalitarian enslavement by means of a global economic network of computerized bartering. This system will be led by a global dictator whom the Bible calls the *Antichrist.* This Antichrist, Satan's personal representative on the scene—the final "Hitler"—will successfully orchestrate the affairs of the entire globe via a One World Government. This man will be the very epitome of evil. The Bible tells us about the number **666** (of which Hollywood has made such a big deal in horror films). Of course, as is typical with Hollywood, the number *really is* a "big deal," even though they treat it as fiction or a myth. Consider the following: "That no man might buy or sell, save he that had the mark, or the name of the beast, or the number of his name. Here is wisdom. Let him that hath understanding count the number of the beast: for it is the number of a man; and his number is Six hundred threescore and six" (Revelation 13:17–18).

As we are admonished to seek understanding and wisdom, for generations, Bible scholars have been studying this mystery in an attempt to identify who would become this world leader, this Antichrist. It is not such an impossible task as it may seem, because the New Testament was written in Greek, and Greek characters not only comprise their alphabet, but each character carries a numerical value as well. If you were a good student of Greek and knew your Scriptures well, you could become very intrigued with checking the names of world leaders who are volleying for position on the international horizon today. However, even though it seems impossible that this future dictator is not now alive somewhere, the Bible says that he will not be revealed for sure until about halfway through the Great Tribulation. If, indeed, we *are* the generation who will witness the return of Jesus Christ, then this leader surely is born at this time and working his way up through the ranks—although whoever it is may come out of seeming obscurity and rise quickly to power because of the abilities given to him by Satan.

The Bible is very clear on one thing, however—that even the very elect will be deceived if they are not alert and do not exercise discernment.

The 666 "Mark" Economic System

In the Antichrist's global economic system, *no one* will be able to work, eat, own property, or "buy and sell" anything without accepting a world-government "Mark" (perhaps a biochip) in his or her right hand or forehead. This New World Order biochip Mark will be the means by which all financial transactions are consummated. Without it, no one will be able to transact any business anywhere in this forthcoming *cashless, debit, computerized, bartering system.*

This sounds very sinister, doesn't it? Like science fiction? Like your worst nightmare? Well, it is, but the worst is yet to come. If one refuses to cooperate with this global program of electronic bondage by refusing the Mark, he will be killed. (In Revelation, the method is beheading, and the relevance of the method will become apparent in future chapters.) On the other hand, if one *does* accept the Mark, he will burn in Hell forever, according to Scripture (Revelation 14:9–11).

We are told in the book of Revelation that whoever resists taking the Mark will be coerced into compliance under the threat of death.

For rejecting it, he will lose his head.

How horrible. Could this be true? Yes, God tells us that this will be a classic "do or die" situation that will be the final test of all ages. Will you worship Jesus Christ or the Antichrist? There is no middle ground—either you are *for* Jesus Christ or *against* Him. If you are not *for* Him, by virtue of that decision, you automatically are *against* Him. It's that simple.

The Satanic New World Order

It's been assumed for centuries that a prerequisite for the coming of Antichrist would be a "revived" world order—an umbrella under which national boundaries dissolve and ethnic groups, ideologies, religions, and economics from around the world orchestrate a single and dominant sovereignty. At the head of the utopian administration, a single personality will surface. He will appear to be a man of distinguished character, but will ultimately become "a king of fierce countenance" (Daniel 8:23). With imperious decree, he will facilitate a One World Government, universal religion, and global socialism. Those who refuse his New World Order will inevitably be imprisoned or destroyed until, at last, he exalts himself "above all that is called God, or that is worshipped, so that he, as God, sitteth in the temple of God, showing himself that he is God" (2 Thessalonians 2:4).

For many years, the notion of an Orwellian society where a One World Government oversees the smallest details of our lives and in which human liberties are abandoned was considered anathema. The idea that rugged individualism would somehow be sacrificed for an anesthetized universal harmony was repudiated by America's greatest minds. Then, in the 1970s, things began to change. Following a call by Nelson Rockefeller for the creation of a "New World Order," presidential candidate Jimmy Carter campaigned, saying, "We must replace balance of power politics with world order politics." This struck a chord with international leaders, including President George Herbert Walker Bush, who in the 1980s began championing the one-world dirge, announcing over national television that the time for a "New World Order" had arrived. The invasion into Kuwait by Iraq/Babylon provided perfect cover for allied forces to engage the Babylonian "prince" by launching Desert Storm against Saddam Hussein's forces, an effort Bush made clear was "to forge for ourselves and for future generations a New World Order...in which a credible United Nations can use its...role to fulfill the promise and vision of the UN's founders." Following this initial statement, Bush addressed the Congress, adding:

What is at stake is more than one small country [Kuwait], it is a big idea—a New World Order, where diverse nations are drawn together in common cause to achieve the universal aspirations of mankind.... Such is a world worthy of our struggle, and worthy of our children's future...the long-held promise of a New World Order. (President George H. W. Bush, address before Joint Session of Congress on the State of the Union, January 29, 1991)

Ever since the president's astonishing newscast, the parade of political and religious leaders in the United States and abroad pushing for a New World Order has multiplied. Britain's Prime Minister Tony Blair, in a speech delivered in Chicago, April 22, 1999, said frankly, "We are all internationalists now, whether we like it or not." Blair could barely have imagined how quickly his doctrine would catch on. By December 9, 2008, respected chief foreign affairs columnist for *The Financial Times*, Gideon Rachman (who attended the 2003 and 2004 Bilderberg meetings at Versailles, France, and Stresa, Italy), admitted, "I have never believed that there is a secret United Nations plot to take over the US. I have never seen black helicopters hovering in the sky above Montana. But, for the first time in my life, I think the formation of some sort of world government is plausible." The United Kingdom's Gordon Brown not only agreed, but in an article for *The Sunday Times*, March 1, 2009, said it was time "for all countries of the world" to renounce "protectionism" and to participate in a new "international" system of banking and regulations "to shape the twenty-first century as the first century of a truly global society." On January 1, 2009, Mikhail

Gorbachev, the former head of state of the USSR, said the global clamor for change and the election of Barack Obama was the catalyst that might finally convince the world of the need for global government. In an article for the *International Herald Tribune*, he said:

> Throughout the world, there is a clamor for change. That desire was evident in November, in an event that could become both a symbol of this need for change and a real catalyst for that change. Given the special role the United States continues to play in the world, the election of Barack Obama could have consequences that go far beyond that country....
>
> If current ideas for reforming the world's financial and economic institutions are consistently implemented, that would suggest we are finally beginning to understand the important of global governance.

Four days later, on January 5, 2009, the chorus call for a New World Order was ramped up by former Secretary of State Henry Kissinger while on the floor of the New York Stock Exchange. A reporter for CNBC asked Kissinger what he thought Barack Obama's first actions as president should be in light of the global financial crises. He answered, "I think that his task will be to develop an overall strategy for America in this period, when really a New World Order can be created." Kissinger followed on January 13 with an opinion piece distributed by Tribune Media Services titled, "The Chance for a New World Order." Addressing the international financial crises inherited by Barack Obama, Kissinger discussed the need for an international political order (world government) to arise and govern a new international monetary and trade system. "The nadir of the existing international financial system coincides with simultaneous political crises around the globe," he wrote. "The alternative to a new international order is chaos." Kissinger went on to highlight Obama's extraordinary impact on the "imagination of humanity," calling it, "an important element in shaping a New World Order" (Henry Kissinger, "The Chance for a New World Order," *Real Clear Politics*, January 13, 2009, http://www.realclearpolitics.com/articles/2009/01/the_chance_for_a_new_world_ord.html). Kissinger—a Rockefeller functionary and member of the Bilderberg group and Trilateral Commission who routinely turns up in lists among senior members of the Illuminati—peppered his article with key phrases from Masonic dogma, including the comment about the "alternative to a new international order is chaos," a clear reference to *ordo ab chao* from ancient Craft Masonry, a reference to the doctrine of "order out of chaos." Like the mythical phoenix firebird, Kissinger visualized the opportunity for a New World Order to be engineered from the ashes of current global chaos, exactly the point he had made years earlier at the Bilderberger meeting in Evian, France, on May 21, 1991, when describing how the world could be manipulated into willingly embracing a global government. He said:

> Today Americans would be outraged if UN troops entered Los Angeles to restore order; tomorrow they will be grateful. This is especially true if they were told there was an outside threat from beyond, whether real or promulgated, that

threatened our very existence. It is then that all peoples of the world will plead with world leaders to deliver them from this evil. The one thing every man fears is the unknown. When presented with this scenario, individual rights will be willingly relinquished for the guarantee of their well being granted to them by their world government. (Transcribed from a tape recording made by one of the Swiss delegates.)

During his second inaugural address, US President George W. Bush likewise envisioned the specter of a Babylonian-like, One World Government. With an almost religious tone, he cited Masonic script, saying, "When our Founders declared a new order of the ages…they were acting on an ancient hope that is meant to be fulfilled" (President George W. Bush, second inaugural address, January 20, 2005).

New Age guru Benjamin Creme was clearer still on how the marriage of politics and religion would epitomize the New World Order when he said some years ago, "What is the plan? It includes the installation of a new world government and a new world religion under Maitreia [a New Age "messiah"]" (Pat Robertson, *The New World Order*, Dallas: Word, 1991, 5). Five-time United States senator from Arizona and Republican presidential nominee in 1964, Barry Goldwater, likewise foresaw the union of politics and religion as a catalyst for global government. In writing of the efforts of behind-the-scenes groups, including international bankers, to bring about a New World Order, he said it would occur through consolidating "the four centers of power—political, monetary, intellectual, and ecclesiastical." As the managers and creators of the new (prophetic) system, this power elite would "rule the future" of mankind, he believed (Barry M. Goldwater, *With No Apologies: The Personal and Political Memoirs of United States Senator Barry M. Goldwater*, 1st ed., New York: Morrow, 1979, 284). So concerned was Goldwater with the consolidation of government policy and religious creed that on September 16, 1981, he took the unique position of warning political preachers from the floor of the US Senate that he would "fight them every step of the way if they [tried] to dictate their [religious ideas] to all Americans in the name of conservatism." The increasing influence of the Religious Right on the Republican Party was bothersome to Goldwater in particular because of his libertarian views. It should have concerned theologians as well, and I say this as a man often associated with the Religious Right. Combining religious faith with politics as a legislative system of governance hearkens the formula upon which Antichrist will come to power. (Note how in the book of Revelation, chapter 13, the *political* figure of Antichrist derives ultra-national dominance from the world's *religious* faithful through the influence of an ecclesiastical leader known as the False Prophet.)

Neither Jesus nor His disciples (who turned the world upside down through preaching the gospel of Christ, the true "power of God," according to Paul) ever imagined the goal of changing the world through supplanting secular government with an authoritarian theocracy. In fact, Jesus made it clear that His followers would not fight earthly authorities purely because His kingdom was "not of this world" (John 18:36). While every modern citizen—religious and non-religious—has responsibility to lobby for moral good, combining the mission of the church with political aspirations is not only unprecedented in New Testament theology—including the life of Christ and the pattern of the New Testament church—but, as Goldwater may have feared, a tragic scheme concocted by sinister forces to defer the church from its true power while enriching insincere bureaucrats, a disastrous fact that only now some are beginning to understand.

Behind these scenes and beyond view of the world's uninitiated members, the alchemy and rituals of the occult masters—Illuminatists, Masons, Bonesmen, Bilderbergers, and Bohemians—have combined to harmonize so completely within recent US foreign and domestic policies as to clearly point to a terrifying Sibyl's conjure, a near-future horizon upon which a leader of indescribable brutality will appear. Although this false prince of peace will seem at first to hold unique answers to life's most challenging questions, ultimately he will make the combined depravities of Antiochus Epiphanes, Hitler, Stalin, and Genghis Khan, all of whom were types of the Antichrist, look like child's play. He will raise his fist, "speaking great things…in blasphemy against God, to blaspheme his name, and his tabernacle, and them that dwell in heaven" (Revelation 13:5–6). He will champion worship of the "old gods" and "cause that as many as would not worship the image of the beast should be killed" (Revelation 13:15), and he will revive an ancient mystery religion that is "the habitation of devils, and the hold of every foul spirit, and a cage of every unclean and hateful bird" (Revelation 18:2).

Nevertheless, the world is readied—indeed, hungry for—a political savior to arise now with a plan to deliver mankind from upheaval.

Loss of Privacy and Electronic Bondage

E ven as we write this chapter, we are reminded of the horrendous events that unfolded in Oklahoma City in April 1995, when terrorists bombed a nine-story federal building, allegedly with a single truck bomb parked in front of the structure. Supposedly, the terrorists just parked the truck, drove away in another vehicle, and detonated the massive bomb from a remote location, or perhaps with a timing device. (Currently, the feasibility of this method is being challenged by some who are experts on the construction of this particular building. Their feeling is that additional internal explosives would have been required, and placed on certain crucial supports, to have rendered the kind of damage that resulted.)

This terrorist act affected the federal building, which housed primarily what would be referred to as civilian activities, i.e., Social Security offices, federal assistance offices, and a sizable daycare center, as well as a few such as FBI, CIA, and Secret Service, but the devastation was not limited to that one building. The shock was felt as far away as fifty miles, and the glass in buildings as far away as four to five blocks was blown out, injuring many unsuspecting pedestrians and office workers, as well as a number of other children in the YMCA facility across the street.

The fatalities surpassed one hundred fifty, with a few bodies never located. The medical personnel ran out of body bags and requested that people donate sheets. Dogs with high-tech, extremely sensitive microphones were sent through the rubble in an attempt to locate any remaining survivors. One victim had a leg amputated on the spot in order to be extricated from the debris and have her life saved.

The nine floors "pancaked" down on top of one another and landed in a crater created by the bomb. Despite all the efforts of the engineers to shore up the wreckage, and despite the extreme care taken by the rescue workers, one rescuer was killed and two others were injured in the heroic attempts. The devastation has been compared in magnitude with the earthquake of two years earlier in Northridge, California.

However, there is a big difference between the two events… and the resulting effect on the population both in Oklahoma and across our nation is much more explosive. This act of violence was committed by radical people with a religious or political agenda… hard-line fanatics who practice the principle that the end justifies the means—regardless of the innocent victims who were in no way involved in their cause, either pro or con.

Since this has been hashed and rehashed on every TV station, newspaper, magazine, talk show, and in books, you may be asking yourself why we're bothering to include it here. When this kind of terrorism strikes the heartland of America, rather than its big cities or Washington, DC (where people assume such actions may occur), fear grips the hearts of everyone, and *prevention* becomes the top priority in our minds. A groundswell of cries for justice and protection starts to rise from the grassroots.

Why did this happen? *It could be for any reason in any distorted mind.* Will it happen again? *Undoubtedly.* Can we stop it? *Probably not.* What are we willing to do to *try* to stop it? *Bingo. You are willing to give up some more of your freedoms and willing to submit to more covert surveillance and more control of your lives...because how else can we protect the safety of all those innocent victims?*

Of course, the ultimate example of this social phenomenon has increased exponentially since September 11, 2001. The subject has since been discussed and rehearsed ad nauseam with the conclusion that we need to increase our surveillance techniques and strengthen our laws to aid law enforcement units in the tracking and capturing of dangerous dissidents. It has already been pointed out that the CIA, FBI, and Secret Service should have better information-gathering abilities, so they can know about these events before they happen and arrest the culprits on some kind of conspiracy charge before a criminal act has actually been committed.

Everything that's in us wants to shout, "Yes!" But that brings us back to the question of how much liberty and freedom we're willing to forego. Are we going to *invite* "Big Brother" to turn this country into an armed camp (someone on the news referred to it as the "bunker mentality"), with all rights of privacy tossed out the window in the process? If we give over that kind of power, will it stop the terrorism? *No.* Did it stop the IRA from bombing London on a regular basis? Has it stopped the Islamic terrorists from bombing Israel on a regular basis? Did it protect our Marine base from a terrorist bombing? *No, to all of the above.* The very nature of terrorism means that you can't stop it. Terrorists don't operate by the rules, and they are willing, if necessary, to die to deliver the explosives to the target…and they don't care who dies with them.

We might succeed in making them go a little farther undercover, and their actions may become more covert, but if they are crazy enough and determined enough, they'll find a way.

So, what will be accomplished by forfeiting our liberties? We will give the government the right to meddle even more into our businesses and lives…for our own good, of course, as with all the other things it is doing to us for "our benefit."

In addition to adding personnel to law enforcement investigative agencies, the government wants to enhance the ability of these agencies in its electronic eavesdropping, via emails, telephones, fax transmissions, etc. Also, Big Brother wants to help the telecommunications companies pay for the mandated purchase and installation of new software and equipment to provide access to their digital systems, which would allow the government to intercept and monitor all telecommunications—in many cases without a warrant. Because prior to the Oklahoma City bombing and 9/11, the government was trying to force the telecommunications companies to pay for all these expensive conversions themselves, the companies were trying to avoid cooperating.

As far back as 1994, a particularly enlightening article titled "Privacy in the Digital Age" by Bill Machrone for *PC Magazine* pointed out, among other hair-raising information, that the "pending legislation provides for fines up to $10,000 a day against telecommunications companies who don't give the Feds the access they want to the decoded data streams."

The loss of our privacy and electronic bondage go hand in hand. Once you are in the system, it is impossible to extricate yourself from it. And now the various agencies, i.e., social services, health care, insurance systems, medical groups, retail sales, credit bureaus, Social Security, the Internal Revenue Service (IRS), vehicle/driver's license registration, voter registration, etc., are sharing the information in their files about you.

As much as you might like to avoid it, there are plenty of places you probably already exist in Big Brother's "big brain." One of the biggest cracks in the security of your privacy is the rampant use of your Social Security number to currently identify you just about everywhere. The fact that many of the people who use it to identify you freely print it in public view—or make it easily accessible to anyone who seeks it—makes it relatively easy to obtain information on your most personal activities, and even to make changes in your file.

Then there's the IRS. While its mission statement says it exists to "enforce the law with integrity and fairness to all," we've all felt for quite some time that its staffers know more about us than they need to. This was thoroughly illustrated in 2013, following a series of eye-opening developments that led to the resignation of the IRS commissioner and a second IRS official pleading the Fifth Amendment when refusing to answer questions before Congress. The fuss followed the discovery that the agency had been inappropriately harassing conservative organizations—such as veterans groups and the Tea Party movement—and selectively investigating them, demanding private details about their religious preferences, and intentionally delaying applications for tax-exempt status. More recently, the Thomas More Society has filed multiple lawsuits to stop the IRS from also singling out pro-life groups with intrusive questions including, "How often do you pray?"

Before these recent events, the *Los Angeles Times* reported:

> If you have a back tax bill with the Internal Revenue Service, watch out. In the midst of a program called economic reality, the federal tax agency is going on line, searching for signs of noncompliance as well as electronic records of cars, credit and real estate it can seize from delinquent taxpayers.... A cadre of IRS agents with computers and modems now will be searching records filed with Department of Motor Vehicles, county tax assessor's offices, credit-reporting companies and the U.S. Bureau of the Census in an effort to find people who are underreporting their business sales, overestimating their deductions or trying to hide assets—or themselves—from federal tax collectors. IRS officials say... "We will be using information from various [electronic] sources as part of our economic-reality approach"... The IRS will begin compiling a host of demographic information about people in each district.... This information will include currency and banking reports, license information, construction contact information and census data.... The IRS will get current addresses for taxpayers who have apparently

dropped off the rolls by buying them from credit-reporting companies.... It can get your full credit file to determine whether you have enough credit to pay the bill.... DMV records will be tapped...to see if you have a car to sell to pay taxes—and to help determine whether a taxpayer is lying about income or deductions. The IRS will be suspicious, for example, of a waiter who reports $20,000 in total income but drives a new Porsche. Property records will be used in the same way.... A few credit experts warn that it also puts a burden on individuals who are under IRS scrutiny. Why? The records are not always right. And *the tax agency does not need to inform you that it is searching these records, nor is it required to allow you to correct records that are in error*...it cannot correct somebody else's database.... If you're under IRS scrutiny, it may behoove you to check your own records for accuracy.

Another article on the IRS from the *Reno Gazette-Journal* discussed how IRS auditors have become "gumshoes" and are training auditors—formerly accountants, for the most part—to be detectives, using all databases at their disposal to compile a composite of YOU—what you should be driving and where you should be living (based on the income you reported), weddings of your children, your cultural background, vacations, home furnishings, etc. They want to develop a complete profile on you. And these techniques have now become standard practice for *all audits*.

The plan to build such databases for monitoring our activity and to combine such Big Brother bureaucracy with a national ID system has been quietly under development for decades. Consider how over twenty years ago *Spotlight* magazine (June 13, 1994) warned under the subtitle, "Danger in the Mail…Now that I Have Free Access to Your Bank Records…Your House Is Next. Nibbling Away Your Freedom Bit by Bit."

Don't look now, but Uncle Sam has some shiny new shackles with your name on them. Indeed, sources in the US Postal Service recently revealed that they're all set to deliver your very own personalized federal ball and chain directly to your mailbox.

[The Postal Service] told several people that they were prepared to mail 100 million of the cards in a matter of months.…

The Clinton administration, which says it's determined to "break the cycle of dependency" among welfare recipients, is preparing to reduce every American to total dependence—and near-total surveillance—through these infamous cards.…

The Postal Service's proposal (which was echoed by the IRS—what a coincidence) calls for the card to "mediate" the information about you in every government database. It will be like a magic key, which opens every government database with information about you.

And here's another troubling fact. If federal computer systems are already integrated to this extent—where one card can "unlock" every piece of information about you—then what makes you think you have the only key?

Of course you won't have the only key. And potentially everything you own and all your assets, benefits and entitlements can be "withheld" from you with the push of a few buttons at the Treasury Department, IRS, or who-knows-where....

The databases are ready to be integrated under the card.

The article above reminds of the way Big Brother (in George Orwell's book, *1984*) was able to successfully maintain control over the lives of the people by abolishing all personal privacy. They were under total surveillance, stripped not only of their privacy, but of their freedom, worth, and dignity, as well. Orwell describes the scene this way:

The telescreen received and transmitted simultaneously. Any sound that Winston made, above the level of a very low whisper, would be picked up by it; moreover, as long as he remained within the field of vision which the metal plaque commanded, he could be seen as well as heard. There was, of course, no way of knowing whether you were being watched at any given moment.

How often, or on what system, the Thought Police plugged in on any individual was just guesswork. It was even conceivable that they watched everybody all the time. But, at any rate, they could plug in your wire whenever they wanted to. You had to live—did live, from habit that became instinct, with the assumption that every sound you made was heard, and except in darkness, every move was scrutinized.

The Clinton administration consistently pushed for a number of new, high-tech systems to enhance their "people control and monitoring" activities. Among their favorites: the national ID card, the Information Superhighway, "and installation of a federal 'Clipper chip' in our telephones, computers, [and modems], fax machines, and other electronic devices," to allow the government easy access for the purpose of tapping and monitoring *all* of our communications via those systems. Don McAlvany informs us that even though the clipper chip project was pushed hard by Clinton, Janet Reno, and former FBI director Louis Freeh, it actually was launched by George Bush in 1991 and developed by the National Security Agency (NSA), a supersecret organization. The August 1994 issue of *The McAlvany Intelligence Advisor* reports as follows:

> Reno and Freeh are presently pushing Congress to enact requirements that telecommunications providers (i.e., local telephone services, cellular phone companies, wireless services, long distance networks, etc.) be mandated to develop and install software and equipment that allows the government to intercept and monitor *all* telecommunications in America. Freeh and Reno argue that "*to stop terrorism and organized crime, the American people must give up some of their personal freedom and privacy.*" [Note: You can see from the date of this newsletter that this plan was on their agenda long before the Oklahoma City bombing occurred. This just gave them the excuse they needed to push such legis-

lation through the Congress. Kind of makes you wonder whose "plans" benefited most from that act of terrorism.]

The FBI has reintroduced its 1992 proposal to require that communications service providers redesign their equipment to facilitate electronic surveillance. The Digital Telephony and Communications Privacy Improvement Act of 1994 [Note: That title is Orwellian "doublespeak," because the act will *destroy all privacy*.] mandates that phones, cable, and computer network companies modify their switches and computers *to ensure that surveillance can be conducted concurrently from a remote government facility. All transactions and phone calls (in and out) will be monitored and recorded. Companies who refuse to comply will be fined $10,000 per day.*

The Electronic Frontier Foundation has warned: *"The FBI scheme would turn the data superhighway into a national surveillance network of staggering proportions."*

The Clintonistas have said that within a few years they plan to link every home, business, lab, classroom, and library via their high-tech computerized information superhighway. The *Wall Street Journal* warned in an editorial (7/10/94) that even if the Congress blocked the government-backed installation of the "Clipper Chip" that the bureaucracy would make an end run and install it anyway.

The *Wall Street Journal* pointed out that *sophisticated terrorists and organized crime syndicates could easily evade the "Clipper Chip" surveillance, but that it would enable the Big Brother bureaucracy to monitor every phone call, every credit card purchase, every bank transaction, and every telecommunication of every private citizen in America.* The *Wall Street Journal* concluded that: *"The potential for government manipulation and intimidation of the citizenry is enormous."*

USA Today (7/20/94) carried a front-page story entitled "Privacy Abuse Confirms the Worst Fear," which discussed how IRS officials had admitted at a Senate hearing that more than 1,300 IRS agents have been investigated over the past five years on suspicion of improperly snooping through taxpayers' files. About 56,000 IRS employees (nearly half of the agency's 115,000 work force) have access to the Integrated Data Retrieval System (IDRS), the computer system that handles collection and storage of taxpayer information.

Evidence of IRS privacy abuses was revealed by the Senate Governmental Affairs Committee in August 1993. Committee member Senator David Pryor (D–AR) said: "The IRS' disregard of taxpayer (privacy) rights confirms the worst fears that the American people have about the IRS. This illegal and offensive activity must stop, and it's clear that Congress must act." *But the major problem with the IRS violation of taxpayer privacy rights is not just IRS snooping of taxpayer returns. It is the sharing of that information with dozens of other government agencies—a practice which until recently was strictly forbidden.*

The most ominous part of the *USA Today* article was the revelation that "the IRS is in the middle of an **$8 billion** computer systems upgrade. Eventually, optical character readers will be used to scan and direct tax returns into three main computers hooked together in a national network that links the ten IRS service centers, eight regional offices, and sixty-five district offices." (emphasis added)

Because of grassroots pressure and their success in enlightening the populace as to the true meaning and end results of some of these measures, many of the "control" elements of the New World Order have not yet been passed by Congress. But that doesn't seem to be hindering the progress of the current administration as it moves us ever forward toward the New World Order.

No problem. The administration will just enact what it wants via an "executive order." "Executive orders give a president the ability to declare a state of emergency and martial law. This suspends all constitutional rights, converting our democratic form of government into a [total] dictatorship [with merely the stroke of a pen]" (see: http://www.unitedstates1.netfirms.com). Congress can subsequently accept these executive orders, publish them in the *Federal Register*, and establish them as laws of the land. These can be implemented at the whim of the current president on a moment's notice just by declaring a state of emergency.

For example, Clinton used this method to move US troops under UN command (PDD-25 signed May 5, 1994). Executive orders date back as far as Franklin Delano Roosevelt. President Carter signed No. 12148 delegating the power to run the entire country to FEMA (Federal Emergency Management Association), then on June 3, 1994, Clinton signed a new executive order transferring control of the country in an emergency from FEMA to the National Security Council and the national security advisor. Some of the executive orders that subsequently have made their way into the *Federal Register* (and are now laws) are reported by McAlvany as follows:

10995—All communications media seized by the federal government. 10997—Seizure of all electrical power, fuels, including gasoline and minerals. 10998—Seizure of all food resources, farms and farm equipment. 10999—Seizure of all kinds of transportation, including your personal car, and control of all highways and seaports. 11000—Seizure of all civilians for work under federal supervision. 11001—Federal takeover of all health, education, and welfare. 11002—Postmaster General empowered to register every man, woman and child in the U.S.A. 11003—Seizure of all aircraft and airports by the federal government. 11004—Housing and Finance Authority may shift population from one locality to another. Complete integration. 11005—Seizure of railroads, inland waterways, and storage facilities. 11051—The director of the Office of Emergency Planning authorized to put Executive Orders into effect in "times of increased international tension or financial crisis." He is also to perform such additional functions as the president may direct.

In short, if there should be nationwide riots (*a la* Los Angeles in '92) for any reason; a national financial crisis; massive social upheaval (i.e., a huge quantum jump in crime); major resistance to national gun confiscation or to the installation of the New World Order or other socialist/police state measures; etc., *[the president]... and his "com-*

rades" have the power and machinery to instantly suspend the Constitution and declare a total dictatorship. (emphasis added)

On April 19, 1995, an article appeared in the Clifton, New Jersey, *Herald & News*. Writer Rich Calder titled the article, "Clifton Group OKs Listing of Residents; City Council Urged to Register Everyone." This city called for a computer database with which it could *track all residents*.

A 26-member committee…recommended last night that the city council approve an ordinance requiring all residents to register with the city. This proposal would obligate all renters and homeowners to fill out a dwelling certificate… listing all occupants in their household. The certificate would require residents to list their names and ages, along with the names and ages of their children. This measure… may raise constitutional issues. Members of the ACLU have said that this procedure will threaten residents' privacy. Committee members who support the proposal said that people with nothing to hide shouldn't have anything to worry about.… Mayor James Anzaldi hailed the plan at last night's meeting, saying the council should implement an ordinance right away.… The dwelling certificate would be supplied by several city agencies.… The proposal also calls for a centralized computer database to be created that would help track…violations.

Have we lost our privacy? For sure. But as one lady put it earlier, "You ain't seen nothin' yet." Are we in electronic bondage? You bet. And we are probably already too far into the system to ever get out. All we can do is sound the warning and get prepared for the battle cry, because for most of us, it's too late to pull a "disappearing act" from the system, unless it's the disappearing act described in the Bible when the Lord returns to snatch us away. However, we don't know exactly when that will be, so we *must be prepared* to survive (victoriously, I hope) in the troubled times that lie ahead, as Scripture is emphatic about one thing: in the end times, things are going to get *worse* before they get better.

But we are not to despair—we have a blessed hope and lots of promises in the Word from our heavenly Father, such as He "will never leave us nor forsake us," His "strength is made perfect in our weakness," "I've never seen the righteous forsaken, or their seed out begging for bread," and, finally, His instruction to "comfort one another with these words": "When you see these things begin to come to pass, look up for your redemption draweth nigh."

chapter three

NWO Organizations—The Illuminati and Friends

The Council on Foreign Relations (CFR), Trilateral Commission (TLC), and Bilderbergers (with covert assistance from others) are said to run the world through manipulation of the banking systems. In the United States it is the Federal Reserve, a private organization that most Americans *incorrectly assume* is owned, run, and/or controlled by our government. In other countries, the World Bank is in control of international finance.

The Illuminati

Webster's Illustrated Encyclopedic Dictionary gives the following definition under the word "illuminati," and it couldn't be more accurate if it had been written by your traditional right-wing conservative.

1) Persons claiming to be unusually enlightened with regard to some subject. 2) a. The members of a secret society of freethinkers and republicans that flourished in Germany during the late 18th century. Also called "Illuminaten." b. Persons regarded as atheists, libertines, or radical republicans during the 18th century (such as French Encyclopedists, the Freemasons, or the freethinkers). 3) The members of a heretical sect of 16th-century Spain, who claimed special religious enlightenment.

Consider the *first* definition: "Persons *claiming* to be unusually enlightened with regard to some subject." Members of the Illuminati consider themselves to be the only ones qualified and sufficiently enlightened to run the world, and they are power-hungry enough to scheme until they actually control it...and us. The *second* definition (part "a") refers to a "secret society" going back to Germany as far as the eighteenth century; part "b" says they were regarded as atheists, libertines, and radicals, and further adds a reference to the Freemasons. The *third* brings in the religious aspect of the Illuminati, calling it a heretical sect claiming special religious enlightenment.

Where the CFR, Trilateral Commission, and Bilderbergers emphasize control of the world through control of world finances, the Illuminati/Freemasons, Skull & Bones, and similar secret societies found on university campuses around the world focus on education, or the control of the minds of our future leaders. Of course, they all adhere heavily on the "good ol' boy" network, where anyone who belongs to the organization is assured successful placement in a position of power and prominence. In personnel or other selection, priority is always given to candidates who are brothers of this elite group over applicants or candidates who are not.

Antony Sutton, former research fellow at the Hoover Institution, Stanford University, as well as professor at California State University, Los Angeles, has authored a book entitled *America's Secret Establishment: An Introduction to the Order of Skull & Bones.* It details the Order of the Skull & Bone (without becoming an initiate), gives a brief background on the origin of the Illuminati, and makes the case for a plausible link between the two groups, though the author is careful to point out that his documentation is inconclusive as yet.

> The Illuminati [a secret society] was a group of Bavarian conspirators dedicated to the overthrow of government. The society was founded on May 1, 1776, by Adam Weishaupt, Professor of Canon Law at the University of Igolstadt [while in America we were busy drafting the Declaration of Independence]. The Order of the Illuminati presumably ceased to exist when it was raided by the Bavarian police in 1786.... The Order was dissolved and its seized papers published. Because the Bavarian state ordered the Illuminati papers published, we have authentic information about the organization and its methods of operation.... Subsequent investigation [of those documents] determined that the aim of the Illuminati was world domination, using any methods to advance the objective, i.e., the end always justifies the means. It was anti-Christian although

clergymen were found in the organization. Each member had a pseudonym to disguise his identity [a truly secret society in every sense of the term]. (See: Antony Sutton, "America's Secret Establishment—An Introduction to Skull and Bones," *American Buddha Online Library*, last accessed September 25, 2013, http://www.american-buddha.com/cia.americasecretestab.1.3.htm.)

The Illuminati's concept of education can be traced to the influence of early nineteenth-century German philosophers. These concepts were introduced in the United States by postgraduate students studying in Europe, bringing their ideas back home with them, then instigating the Illuminati plan to educate our youngsters according to their goals and philosophies.

Antony Sutton states (quoting John Robinson in *Proofs of a Conspiracy*): "So far as education is concerned, the Illuminati objective was as follows: 'We must win the common people in every corner. This will be obtained chiefly by means of the schools, and by open, hearty behavior, show, condescension, popularity, and toleration of their prejudices *which we shall at leisure root out and dispel* '" (emphasis added; ibid.).

Johann Friedrich Herbart was a major German philosopher when the Yale postgraduate students were studying there. Herbart adhered to the Hegelian philosophy (the state is superior to the individual) and thoroughly indoctrinated his protégés in this teaching.

Therefore, "For Herbart, education had to be presented in a scientifically correct manner, and the chief purpose of education," in his opinion, was:

…to prepare the child to live properly in the social order, of which he is an integral part. The individual is not important. The mere development of individual talent, of individual fitness, mental power, and knowledge is *not* the purpose of education. The purpose is to develop personal character and social morality, and the most important task of the educator is to analyze the activities and duties of men within society. The function of instruction [according to Herbart] is to fulfill these aims and impart to the individual *socially desirable ideas.* (emphasis added; Antony Sutton, "America's Secret Establishment")

In today's vernacular, we would call this being "PC" (politically correct). All these ideas in today's American educational philosophy can be recognized as originating and being transmitted by members (knights and patriarchs) of Skull & Bones, having been learned at the feet of Illuminati educators.

As we progress further into this chapter, you will become increasingly aware of how many people are members of two or more such orders. For example, former president George H. W. Bush is a member of the Order of Skull & Bones (as was his father before him), as well as a member of both the Council on Foreign Relations and the Trilateral Commission. Other prominent contemporary members of Skull & Bones include Winston Lord, former ambassador to China; US senators David Boren of Oklahoma and John Chaffe of Rhode Island; as well as William F. Buckley (allegedly conservative publisher of *The National Review*), among others.

According to Sutton, those on the "inside" of Skull & Bones

know it simply as "The Order." Others have known it for more than one hundred and fifty years as "Chapter 322" of a German secret society. For legal purposes it was incorporated in 1956 as the Russell Trust. It was also once known as the "Brotherhood of Death." The casual name (or the name sometimes used derogatorily) is "Skull & Bones," or just plain "Bones," and it is not just another Greek letter campus fraternity with passwords and secret handshakes. It is far more insidious. Chapter 322 is a *secret* society whose members are sworn to silence (they are supposed to actually leave the room if someone outside the Order even mentions the name Skull & Bones). So far as we can determine, it exists only on the campus of Yale University, though rumors are beginning to surface indicating there may be a select number of other locations; additionally, rumors have begun to circulate regarding possible links to a couple of other secret societies on the Yale campus: the Scroll & Key and Wolf's Head, both founded in the mid-nineteenth century. Allegedly, these are competitive societies; however, Sutton believes them to be part of the same network. It has rules and rituals…ceremonial rites which we will mention briefly a little later.

Sutton states that its members "always deny membership…and in checking hundreds of autobiographical listings for members, [he] found only half a dozen who cited an affiliation with Skull & Bones" (ibid.). He is concerned about whether the many members of the various administrations (either elected or appointed) have declared their membership in the biographical data supplied to the FBI for their obligatory "background checks," implying that it is not likely.

Further, Sutton asserts, then documents, that the Order is unbelievably powerful.

Skull & Bones is an organization of only senior students at Yale University. Each year, only fifteen initiates are selected in their junior year. The Patriarchs (see below) only meet annually on Deer Island in the St. Lawrence River.

Admission to Skull & Bones is by invitation only; there is no lobbying, electioneering, or applying for membership. During commencement week, the juniors are privately "tapped." The junior is given an option: "Skull & Bones. Accept or reject?" This method has not changed since the Order's inception in 1832. "Those who accept, presumably the greater number, are invited to attend the Bones Temple on campus to undergo an initiation ceremony [described briefly below]… For the ambitious, 'tapping' is the magic password to a future [success-guaranteed] career." Potential candidates are apparently selected based on their school and extracurricular activities, their support of Yale, and particularly their sports ability—teamwork is held in very high esteem. The most unlikely choice would be "a loner, an iconoclast, an individualist, [a person who] goes his own way in the world." They want people who put the Order first, without question, and who will abide by the rules at all costs. Sutton states:

> The most likely potential member is from a Bones family, is energetic, resourceful, political, and probably an amoral team player. A man who understands that to get along you have to go along. A man who will sacrifice himself for the good of the team. A moment's reflection illustrates why this is so. In real life, the thrust of the Order is to bring about certain objectives. Honors and financial rewards

are guaranteed by the power of the Order. But the price of these honors and rewards is sacrifice to the common goal, the goal of the Order. Some, perhaps many, have not been willing to pay this price. (Antony Sutton, "America's Secret Establishment"; also see: http://www.mimico-by-the-lake.com)

Initiates undergo bizarre rituals and initiation ceremonies with sexual overtones, conducted on the order of brainwashing techniques designed to strip the initiate of all pride and sense of self. The initiates are then "reprogrammed" to embrace only the philosophies and goals of the Order. (Sutton describes these rituals in great detail in his book.)

At any given time, about five to six hundred members are alive and active. Roughly about one quarter of these take an active role in furthering the objectives of the Order; the others either change their minds or just lose interest and become silent dropouts.

It has been postulated what the number "322" under the skull and crossed bones of the Order's logo or official emblem means. Since the organization was imported from Germany in 1832 as the second chapter of a German order, the "32" denotes the year of origination and the following "2" indicates that it is the second chapter of the organization. A much more likely theory is that "the Order is descended from a Greek fraternal society dating back to Demosthenes in 322 BC.... Bones records are dated by adding 322 [years] to the current year, i.e., records originating in 1950 are dated Anno Demostheni 2272" (ibid).

Inside the Order there are many other numbers and symbols

that appear beside the names of the members in the "Catalogue" (or membership list). Although we don't know for sure, Sutton makes some interesting speculations regarding their possible meaning. The Catalogue contains the birth names of the members, but Sutton tells us: "Entry into the Order is accompanied by an elaborate ritual and no doubt psychological conditioning. For example: Immediately on entering Bones the neophyte's name is changed. He is no longer known by his name as it appears in the college catalogue but like a monk or Knight of Malta or St. John, becomes Knight so-and-so. The old Knights are then known as Patriarch so-and-so. The outside world are known as Gentiles and vandals."

In the century and a half years since the Order was founded, active membership has evolved into a core group of perhaps twenty to thirty families, according to Sutton:

It seems that active members have enough influence to push their sons and relatives into the Order, and there is significant intermarriage among the families. These families fall into two major groups.

First, we find old line American families who arrived on the East Coast in the 1600s, e.g., Whitney, Lord, Phelps, Wadsworth, Allen, Bundy, Adams, and so on.

Second, we find families who acquired wealth in the last 100 years, sent their sons to Yale, and in time became almost old line families, e.g., Harriman, Rockefeller, Payne, Davison.... In the last 150 years, a few families in the Order have gained enormous influence in society and the world. (Antony Sutton, "America's Secret Establishment")

Although the "old line" families used to look down their blue-blood noses at the *nouveau riche*, that is not so much the case anymore. They have discovered it takes money to maintain their power—not just the name—so if they couldn't make it on their own, they "married" it into the family. But whatever it takes, you maintain the power at all costs.

The Order has penetrated every segment of American society, e.g., the White House, legislatures, political parties, foundations (charitable and otherwise), think tanks, policy groups, education, media, publishing, banking, business, industry, commerce, and even churches. In many cases, the Order is the founder of these organizations. It determines the goals and objectives, as well as the methods for achieving them, then usually installs the first president, chairman, or CEO and/or CFO. This explains why so many such organizations are based on the premise of secular humanism. Their roots can be traced directly back to active members of the Order.

Since it seems the most unlikely, let's first take a look at the connection of the Order to the church. Although the percentage has declined in recent years, about two percent of the Order is in the church—all Protestant denominations.

A key penetration is the Union Theological Seminary, affiliated with Columbia University in New York. This Seminary, a past subject of investigation for Communist infiltration, has close links to the Order. Henry Sloane Coffin (1897) as Professor of Practical Theology at Union from 1904 to 1926 and President of Union Theological Seminary, also known as the "Red Seminary," from 1926 to 1945. Union has such a wide interpretation of religious activity that it has, or used to have, an Atheists Club for its students.

Henry Sloane Coffin, Jr. (1949) was one of the Boston Five indicted on federal conspiracy charges.

And this is only part of the Order's penetration into the church. (Antony Sutton, "America's Secret Establishment")

Mr. Sutton expands on the connection between the Order and the other areas, i.e., the law, communications, industry, the Federal Reserve/banking, and the White House/politics/government.

The Law

The major establishment law firms in New York are saturated with the Order.

In particular, Lord, Day, and Lord, dominated by the Lord family already discussed; also, Simpson, Thatcher, and Bartlett, especially the Thatcher family; David, Polk, Wardwell, and Debevoise, Plimpton, the Rockefeller family law firm.

Communications

There has been a significant penetration into communications. Some examples:

- Henry Luce of *Time-Life* is in the Order.
- So is William Buckley ('50) of *National Review.*
- And Alfred Cowles ('31), president of Cowles Communications, *Des Moines Register, Minneapolis Star*

- And Emmert Bates ('32) of Litton Educational Systems, plus
- Richard Ely Danielson ('07) of *Atlantic Monthly*
- Russell Wheeler Davenport ('23), *Fortune*
- John Chipman Farrar ('18) of Farrar, Strauss, the publishers.

The most prestigious award in journalism is a Nieman Fellowship at Harvard University. Over 300 were granted from 1937 to 1968. The FIRST director of the Nieman Fund was member Archibald McLeash.

Industry

The oil companies have their links to the Order. Members Percy, Rockefeller, the Paynes, the Pratts, all link to Standard Oil. Shell Oil, Creole Petroleum, and Socony Vacuum also link. A wide variety of manufacturing firms have members in the Order, from the Donnelly family in Chicago (printers of The Official Airline Guide and other references); lumber companies like Weyerhauser, who is also a Trilateralist; Dresser Industries, and so on.

The Federal Reserve System

A dozen members can be linked to the Federal Reserve, but one appointment is noteworthy, Pierre Jay ('92), whose only claim to fame in 1913 was to run a private school and be an obscure vice president of Manhattan Bank, yet he became first chairman of the New York Federal Reserve, the really significant Federal Reserve.

The White House, Politics, and Government

This is the area where the Order has made headway, with names like Taft, Bush, Stimson, Chafee, Lovett, Whitney, Bundy, and so on. (Antony Sutton, "America's Secret Establishment")

Although we are not yet to our discussion of the CFR and TLC, we briefly address them here as they relate to Skull & Bones Chapter 322.

Though not looking for a spotlight on their discussions and activities, the CFR and TLC are not secret organizations in the sense that the Illuminati and Skull & Bones are secret. Their membership rosters are available to anyone who asks, and they are making the dates and locations of their meetings available to the media, though the media is not permitted inside to witness the activities, nor are members permitted to give interviews revealing what has occurred at the meetings (as spelled out in their bylaws).

"The Order is represented in these organizations, but does not always dominate. David Rockefeller, former chairman of the CFR, is not a member of the Order" (Antony Sutton, "America's Secret Establishment"), but the family was represented in the Order by Percy Rockefeller; however, a later CFR chairman, Winston Lord, is a member of Bones.

Visualize three concentric circles, consisting of the inner core, the inner circle, and the outer circle. The outer circle is made up of large, open organizations (e.g., CFR or TLC) some of whose memberships are made up of Bones members. The inner circle is made up of one or more secret societies, such as Chapter 322. The inner core is believed to be a secret society within the Order, a decision-making core. As yet, the existence of this inner core cannot be documented, but evidence points to its existence. Sutton even believes that he could identify the chairman.

The CFR is the largest organization in the outer circle, with about twenty-nine hundred members at any one time (as many as the Order in its entire history). The TLC has 325 members worldwide, but only about eighty in the United States. Other groups in the outer circle include the Pilgrim Society, the Atlantic Council, the Bilderbergers, and the Bohemian Club (of San Francisco). As an example of the number of Bones members who are also members of the CFR, following are just the Bonesmen whose last names began with the letter "B": "Jonathan Bingham (congressman); William F. Buckley (editor, *National Review* and the Order's house conservative); McGeorge Bundy (foundation executive); William Bundy (Central Intelligence Agency); George Herbert Walker Bush ([former] president of the United States)" (Antony Sutton, "America's Secret Establishment").

The Trilateral Commission was founded in 1973 by David Rockefeller, who was not a member of the Order. The same appears to be true of many TLC members; however, if you investigate the family tree, you will find members of the Order sometimes as close as one generation removed, and in other cases in the family line in great numbers. Sutton points out that the TLC is not a conspiracy, it just doesn't publish its activities too widely, as the general populace probably wouldn't appreciate what it is planning for us behind closed doors.

Sutton reports that he "has openly debated with George Franklin, Jr., coordinator of the Trilateral Commission, on the radio. Mr.

Franklin did show a rather ill-concealed dislike for the assault on his pet global New World Order—and made the mistake of attempting to disguise this objective." Where the TLC is concerned, even though David Rockefeller is not a Bones member, keep these facts in mind: J. Richardson Dilworth, chief financial and administrative officer for the Rockefeller Family Associates, is a member of the Order, as was Percy Rockefeller (1900). And the TLC's purposes, as portrayed in its own literature, are almost identical to those of the Order.

So far as Sutton can determine, William F. Buckley was the only member of both the Order and the Bilderbergers. The Pilgrim Society has no current Bones members, but the family names of past members are found, i.e., Aldrich and Pratt.

We have discussed the guarantees of success for Bones members, as well as the intermarriages to consolidate the power, wealth, and influence of the families. Now, we will look at the chain of influence.

"A chain of influence spread over many years guarantees continuity and must be extraordinarily impressive to any new initiate who doubts the power of the Order" (Cybrarian, "The Germanic Order of Skull and Bones," *CIAgents*, last accessed December 5, 2010, http://www.ciagents.com). "Members of the Order are to be found in every segment of society [or our existence]: in education, foundations, politics, government, industry, law, and finance. Consequently, *at any time the Order can tap influence in any area of society....* [The major occupations, however, are in] law, education, business, finance, and industry" (emphasis added; Antony Sutton, "America's Secret Establishment"). Percentages of Bones members in each area follows: 18 percent, law; 16 percent, education; 16 percent, business; 15 percent, finance; 12 percent, industry. The remaining 23 percent is scattered among all remaining occupations.

Sutton makes a good case for the interference of the Order in promoting wars and conflict in order to maintain and increase its world control: "If we can show that the Order has artificially encouraged and developed revolutionary Marxism national socialism while retaining some control over the nature and degree of the conflict, then it follows *the Order will be able to determine the evolution and nature of the New World Order*" (emphasis added; Antony Sutton, "America's Secret Establishment").

Sutton makes the above statement while documenting a link between the Union Bank and support for the Nazis during World War II. He traced the flow of money through a long trail that attempted to hide its ultimate destination, and proved the involvement of at least eight men, four of whom were members of the Order (including Prescott Bush, father of President George H. W. Bush) and two of whom were Nazis.

As we conclude this section on the Skull & Bones, Brotherhood of Death, I want to discuss the spiritual aspects of the Order. Sutton tells us: "What happens in the initiation process is essentially a variation of brainwashing or encounter group processes. Knights, through heavy peer pressure, become Patriarchs prepared for a life of the exercise of power and continuation of this process into future generations. In brief, the ritual is designed to mold establishment zombies, to ensure continuation of power in the hands of a small select group from one generation to another. But beyond this ritual are aspects notably satanic."

We can make at least three definite statements about links between the Order and satanic beliefs. First, photographic evidence documents the satanic device (as well as the name) of the skull and crossed bones. Second, there is a link to satanic symbolism. Third, the link between the Order and the New World Order is well documented in the book *Hidden Dangers of the Rainbow* by Constance Cumbey.

The skull and crossed bones is not just a logo or printed artwork; photographs exist that show the use of actual skulls and bones in the ceremonies of the Order. According to other evidence, at least three sets of skulls and other assorted human bones are kept within the Bones Temple on the Yale campus. At best, that makes the members grave robbers. But using these bones for ceremonial purposes shows absolutely no respect for the dead, and is a blatantly satanic activity.

Cumbey identified and linked several organizations to the Order and its objectives. She identified Benjamin Creme and the Tara Center as a New Age phenomenon, then linked Creme to the Unity and Unitarian churches. Sutton continued the chain by pointing out the Order's longstanding and significant link to these churches.

"Former president William Taft, whose father co-founded The Order, was President of the Unitarian Association in his time. Cumbey identifies the link between Hitler and the New Age movement and former research by this author linked The Order to the founding and growth of Naziism.... Cumbey states that the New Age movement plans to bring about a New World Order 'which will be a synthesis between the U.S.S.R., Great Britain, and the

United States'" (Antony Sutton, "America's Secret Establishment"). Later information indicates that it will come closer to encompassing the entire world, i.e., both industrialized nations and third-world countries.

"Finally, Cumbey points out that the anti-Christ and satanic aspects are woven into the cult of the New Age movement" (ibid.). The goals and activities of the Skull & Bones and its leaders' plans for our future are undoubtedly satanically inspired.

As we move on to investigate the groups that endeavor to control world governments, wars, etc., through control of finances, we will discuss US money and its emblems. That means going back into the discussion of the Illuminati and Freemasonry, as they are at the root of most of these satanic symbols.

Above you will find the pyramid that is shown on the back of our American currency. Note that the Latin translates: "Announcing the birth of the New World Order."

"Although the pyramid on the American dollar with its thirteen levels ties in with the thirteen colonies, the original association was with ancient Egyptian and Babylonian mysticism.… Note also that the corner stone [or capstone] is missing from the top of the pyramid. In its place is the All-Seeing Eye. [The Illuminati's] mutual spying system was an integral part of [the] program to keep [its] associates in line. The eye symbolized a 'Big Brother' controlling [its] domain. Those who dismiss this idea, say that the eye in the Great Seal is the 'all-seeing' eye of God" ("The Illuminati: Real or Imaginary," *Associate.com*, http://associate.com/ministry_files/ The_Reading_Room/False_Teaching_n_Teachers_3/The_Illuminati_Real_Or_Imagine.shtml). However, the literal translation of the Latin *Annuit Coeptis* and *Novus Ordo Seclorum* indicates quite the contrary: "Announcing the Birth of the New Secular Order," commonly known today as the New World Order.

Before we progress to a cashless society, many changes will take place in our currency. In fact, some have already taken place (under the guise of inhibiting counterfeiting, of course). Former Congressman Ron Paul had this to say about the US Mint facility in Fort Worth (note particularly his comments about the pyramid shape of the building):

The government is crowing, the greenback will be produced on U.S.-made presses for the first time in more than a century. Only it won't be green.

The Stevens Graphics Corporation is producing an ominously named Alexander Hamilton press for the Bureau of Engraving and Printing. (Hamilton was—appropriately enough—a proponent of fiat-paper money, big deficits, and big government)....

The new press can embed plastic or other strips in the bills, do the microprinting the Treasury has talked about, and—oh, yes—print in three colors.

The feds say not to worry. The greenback will remain green. But then why pay for this extra capacity? It can only be to print the New Money.

In an effort to find out more about where the New Money will be printed, I went on an investigative visit to Fort Worth, Texas, to survey a new BEP currency plant. This is no normal federal building. It is one of several places in the country where the New Money is in preparation.

The new BEP building is a monstrosity that perfectly symbolizes unconstitutional abuse of power. A giant windowless blockhouse [circular], it sits on an enormous piece of land, surrounded by a prison-style cyclone fence topped with barbed wire.

The appropriately evil-looking New Money plant went up fast. On my last visit, I noted a new addition to the administrative end of the building: a pyramid. Two interpretations are possible: one, that it copies the Masonic-Illuminatist symbol that also, unfortunately, made it on the back of our one dollar bill; or two, that it symbolizes the kind of society the politicians and bankers have in mind for us: they're the pharaohs and we're the enslaved workers. Or maybe it's both.

Paul pointed out also that former House Speaker Jim Wright and H. Ross Perot were instrumental in the selection of Fort Worth, with the enthusiastic backing of Senator Phil Gramm. He also calls it "a 'public-private partnership' not unlike the Federal Reserve itself."

The Council on Foreign Relations is undoubtedly tied in to all this monetary control through the Federal Reserve and the World Bank, as well as through government influence and other sources of control. In this timeline, we learn a bit about the history of the CFR as it relates to the promotion of the New World Order.

- **May 30, 1919**—Originally founded by prominent British and Americans as two separate organizations: the Royal Institute of International Affairs (in England) and the Institute of International Affairs (in the US). Two years later, Colonel House reorganized the Institute of International Affairs as the CFR.
- **December 15, 1922**—The CFR endorses world government. Philip Kerr, writing for CFR's magazine *Foreign Affairs*, states: "Obviously there is going to be no peace or prosperity for mankind as long as [the Earth] remains divided into 50 or 60 independent states.... Until some kind of international system is created which will put an end to the diplomatic struggles incident to the attempt of every nation to make itself secure.... The real problem today is that of the world government."
- **February 17, 1950**—CFR member James P. Warburg, cofounder of the United World Federalists and son of

Federal Reserve banker Paul Warburg, tells the [Senate Foreign Relations] subcommittee that "studies led me, ten years ago, to the conclusion that the great question of our time is not whether or not one world can be achieved, but whether or not one world can be achieved by peaceful means. We shall have world government, whether or not we like it. The question is only whether world government will be achieved by consent or by conquest."

- **November 25, 1959**—The CFR calls for new international order. "Study Number 7" advocated "a new international order [that] must be responsive to world aspirations for peace, for social and economic change…an international order…including states labeling themselves as 'socialist' [communist]."
- **1975**—In the *New York Times*, CFR member and *Times* editor James Reston writes that President Gerald Ford and Soviet leader Leonid Brezhnev should "forget the past and work together for a new world order."
- **1975**—Retired Navy Admiral Chester Ward, former judge advocate general of the US Navy and former CFR member, in a critique, writes that the goal of the CFR is the "submergence of US sovereignty and national independence into an all-powerful one-world government.… Once the ruling members of the CFR have decided that the US government should adopt a particular policy, the very substantial research facilities of the CFR are put to work to develop arguments,

intellectual and emotional, to support the new policy, and to confound and discredit, intellectually and politically, any opposition."

- **1977**—*Imperial Brain Trust* by Laurence Shoup and William Minter is published. The book takes a critical look at the CFR with chapters titled, "Shaping a New World Order: The Council's Blueprint for Global Hegemony, 1939–1944" and "Toward the 1980s: The Council's Plans for a New World Order."

(Quotes and information from above list can be found at: American Patriot Friends Network, "A History of the New World Order," May 7, 2000, http://www.xld.com/public/cuic/history. htm.)

We are going to "back our way" into this discussion by telling about the CFR, then getting into the Federal Reserve and world financial system…and who controls it. The membership list for the CFR reads like a *Who's Who* of American leaders in every walk of life, i.e., government, private industry, education, the media, military, and high finance.

In 1921, Edward Mandell House founded the CFR (from an organization begun two years earlier, the Institute of International Affairs). He was a close friend and advisor of President Woodrow Wilson. House persuaded the president to support and sign the Federal Reserve Act and to support the League of Nations, forerunner of the UN and other globalist groups. Finances to found the CFR came from the same crowd who formed the Federal Reserve, namely "J. P. Morgan, Bernard Baruch, Otto Kahn, Paul Warburg, and John D. Rockefeller, among others" (H. M. Summers, "The New World Order and the United States of America"; report and more info found here: http://www.angelfire.com/d20/philadelphians/nwo-us.html).

The CFR shares a close cross-membership with other globalist organizations and at the beginning of World War II, the CFR (with the help of Franklin Roosevelt) gained control of the US State Department and, after the war, helped establish the United Nations in 1945. *The US delegation for the UN's founding conference contained forty-seven CFR members.* Those members included "**John Foster Dulles**, **Adlai Stevenson**, **Nelson Rockefeller**, and [Soviet spy] **Alger Hiss**, who was the secretary general of the UN's founding conference" (ibid.).

Many prominent people have served as director of the CFR, including "**Walter Lippmann**, **Adlai Stevenson**, **Cyrus Vance**, **Zbigniew Brzezinski**…**Paul Volcker**…**Lane Kirkland**, **George** [H. W.] **Bush**, **Henry Kissinger**, **David Rockefeller**, **George Shultz**, **Alan Greenspan**, **Brent Scowcroft**, **Jeane Kirkpatrick**, and **Richard B.** [Dick] **Cheney**" (Donald M. Ware, "Transformation: Spiritual, Physical, and Political," *Welcome Truth Seeker*, June 22, 1997; viewable here: http://australiatrade.com.au/Alternative/Space/index.htm?&lang=en_us). Private industry financially supported and/or controlled by the CFR include ARCO, BP, Mercedes-Benz, Seagram, *Newsweek, Reader's Digest, Washington Post*, American Express, Carnegie Corporation, Ford Foundation, GE Foundation, General Motors, Mellon Foundation, Sloane Foundation, Xerox Foundation, IBM, AT&T, Ford, Chrysler, Macy, Federated Department

Stores, Gimbels, Sears, JC Penney, May Dept. Stores, Allied Stores, and various Rockefeller concerns.

The CFR shares many cross-memberships with the Bilderbergers and the Club of Rome, which was founded in 1968 and consists of scientists, educators, economists, humanists, industrialists, and government officials who see it as this organization's task to oversee the regionalization and unification of the entire world. It has divided the world into ten political/economic regions or "kingdoms."

In the Club of Rome we glimpse the dark, spiritual side of the globalist movement. Aurelio Peccei, the Club's founder, revealed in *Mankind at the Turning Point* his pantheist/New Age beliefs, writing about man's communion with nature, the need for a "world consciousness" and "a new and enlightened humanism." He was a student of Pierre Teilhard de Chardin, one of the New Age occultists' most frequently quoted authors.

Club members include the "late **Norman Cousins**, [former] long-time honorary chairman of Planetary Citizens;…**John Naisbitt**, author of *Megatrends*;…**Betty Friedan**, founding president of the National Organization of Women" (ibid.), **Robert Anderson**, and **Harlan Cleveland**, as well as many other New Age speakers and authors. Members also include US congressmen, Planned Parenthood representatives, UN officials, and Carnegie and Rockefeller foundation people.

The CFR has a stranglehold on America. It effectively controls the four most powerful positions (after the presidency) in our government: secretaries of state, treasury, and defense, and the national security advisor.

By the Nixon administration, 115 CFR members held positions in the executive branch. Carter appointed scores of TLC and/or CFR members. All of his National Security Council were or had been members: Mondale, Brzezinski, Vance, Brown, Jones, and Turner. Ronald Reagan appointed 76 CFR and TLC members to key posts, but George Bush broke records with 354 recognizable names in highest levels of government.... The State Department is saturated with CFR/TLC members. Almost every prominent ambassador or diplomat is a globalist.

As for the Treasury Department, Brady and Regan are members, and Federal Reserve chairmen Greenspan, Anderson, Vance, and Volcker are all insiders.

Every US secretary of defense for the past thirty-five years, except Clark Clifford, belonged to either the CFR or the TLC. Every Supreme Allied Commander in Europe and every US ambassador to NATO have been insiders, as well as nine of thirteen CIA directors.

Except for John F. Kennedy and Ronald Reagan, almost every US president was either a CFR or TLC member. JFK may have been a member of the Boston affiliate of the CFR. Ted Kennedy is definitely a member of the affiliate, though his name is not listed in the general CFR membership.

The media is also well represented in the membership of the CFR, including such familiar names as Paley, Rather, Moyers, Brokaw, Chancellor, Levine, Brinkley, Scali, Walters, Schorr, McNeil, Lehrer, and Carter III. Not surprisingly, Rockefellers' Chase Manhattan Bank has minority control of all three major networks (CBS, ABC, and NBC).

The AP, UPI, and Reuters wire services all have CFR members in major positions. Other CFR members in the media include William F. Buckley (also a Bones member), Diane Sawyer, Rowland Evans, and David Gergen.

Most major newspapers today have a strong CFR influence, including the *New York Times, Washington Post, Wall Street Journal, Boston Globe, Baltimore Sun, Chicago Sun-Times, LA Times, Houston Post, Minneapolis Star-Tribune, Arkansas Democrat-Gazette, Des Moines Register & Tribune, The Gannett Co.* (publishes *USA Today* and major newspapers nationwide), *Denver Post*, and the *Louisville Courier*. This is just a partial listing of papers staffed by CFR affiliates.

Magazines with CFR connections include *Fortune, Time, Life, Money, People, Sports Illustrated, Newsweek, Business Week, US News & World Report, Saturday Review, Reader's Digest, Atlantic Monthly, McCall's,* and *Harper's Magazine*.

Book publishers include MacMillan, Random House, Simon & Schuster, McGraw-Hill, Harper, IBM, Xerox, Yale University Press, Little Brown, Viking, Cowles, and Harper & Row. Many of these publish school textbooks, which brings us to just how far the tentacles of CFR influence reach into the education of our children.

CFR foundations, primarily Carnegie and Rockefeller, provided two-thirds of the gifts to all American universities during the first third of last century. And the man who is considered the "father of progressive education," John Dewey, was an atheist who taught four of the five Rockefeller brothers. He spent most of his life educating teachers, including those in the USSR. This man's influence extends not just through years, but throughout generations. Today, 20 percent of all school superintendents within the United States and 40 percent of all education department heads have superior degrees from Columbia, where Dewey headed the education department for many years.

The National Education Association (NEA) adopted his philosophy of humanism, socialism, and globalism, then put it into our classrooms. CFR members head the teaching departments at Columbia, Cornell, New York University, Sarah Lawrence College, Stanford, Yale, University of Chicago, Johns Hopkins University, Brown University, University of Wisconsin, Washington University, and Lee University.

As we have emphasized, this New World Order plan has its roots in the spiritual rather than a material or physical basis, as its proponents would have you believe. Tying the CFR leaders and members directly to the New Age movement ultimately includes the radical ecology groups, Lucis Trust, nominal churches, the UN, and others.

So, how does the New Age fit into the one-world movement? The New Age is pantheistic—the belief that God is the sum total of all that exists. No personal God exists; rather "God" is a force that flows through all living things—and that supposedly makes humans "gods." New Agers believe that since global unity is essential to the proper flow of the god-force, when unity occurs in a One World Government, a new age of enlightenment will emerge. Occult practices (and eastern mysticism) accompany pantheism.

The Theosophical Society (TS) is at the forefront of New Age globalism. Lucis Press (offspring of Lucis Trust, formerly headquartered at the UN) was established by TS leader Alice Bailey. Lucis

Press promotes the preeminence of Satan (Lucifer) and is well connected with the one-world political societies and the World Constitution and Parliament Association. Past and present members include TLC and CFR cross-memberships: "**Robert McNamara, Donald Regan, Henry Kissinger, David Rockefeller, Paul Volcker,** and **George Shultz**" (H. M. Summers, "The New World Order and the United States of America").

The World Constitution and Parliament Association (WCPA) has already written a world constitution, the "Constitution for the Federation of Earth," and has submitted it to world leaders for ratification. Dominated by environmentalists, US personnel, Nobel Laureates, leftist churches (i.e., the WCC), educators, financial leaders, and eastern mystics (pantheists), it calls for an international monetary system; administration of oceans, seabeds, and atmosphere as the common heritage of all humanity; elimination of fossil fuels; redistribution of the world's wealth; complete and rapid disarmament (including confiscation of privately owned weapons); an end to national sovereignty; a global environmental organization; world justice system; and world tax agency.

Many of its members embrace eastern mysticism. United States members include **Jesse Jackson** (also CFR) and former attorney general **Ramsey Clark**. Its director, Philip Iseley, belongs to Amnesty International, the ACLU, Global Education Associates, Friends of the Earth, Sierra Club, Audubon Society, American Humanist Association, SANE (nuclear freeze group), Planetary Citizens, and the Global Futures Network, among many other organizations.

The UN is the chosen agency that is moving us toward world government, but the UN charter is not a *constitution*. Plus, it is an organization of *sovereign* nations, which is anathema to globalists. So the WCPA's world constitution is ready for acceptance at the proper time.

In the November 2, 1994, edition of the *Los Angeles Times*, Doyle McManus writes in an article titled "US Leadership Is More Diverse, Less Influential": "Almost half a century ago, when Harry S. Truman needed help running the foreign policy of the United States at the dawn of the Cold War, the remedy was simple: 'Whenever we needed a man,' one of his aides recalled, 'we thumbed through the roll of Council [on Foreign Relations] members and put in a call to New York.'"

He tells about the expanding of the so-called "elite" to include every area of influence, from academics, scientists, and religious leaders to rock stars and others. Leslie H. Gelb, president emeritus of the CFR, says, "This is the largest foreign policy elite this country has ever enjoyed," and adds that he plans to expand and diversify the "august" organization's membership to embrace new areas such as sports and the arts. "Ironically," says McManus, "just as the elite is growing in size and diversity, it may also be diminishing in power" (ibid.).

Recent surveys, including the *Times Mirror* polls, suggest that the general public is less willing than in the past to accept the advice of the foreign policy elite—at least on issues that come close to home, like free trade with Mexico or the use of American troops in peacekeeping missions overseas.

"There's more information going straight to the general public now," said Brent Scowcroft [CFR and TLC member], who served as national security advisor to President George Bush. "That tends to reduce the influence [of the elite]. It makes it more difficult to get support for a potentially unpopular policy." (Doyle McManus, "US Leadership Is More Diverse, Less Influential," *Los Angeles Times*, November 2, 1994)

The Federal Reserve

Now consider the Federal Reserve, an entity closely tied to the policies and policymakers of the CFR. We all are familiar with the term "prime rate." That is the lowest interest rate set by the Fed (the rate paid by large banks that borrow from them and then increase the interest rate as they lend the money to you, i.e., "prime-plus"). The prime rate fluctuates frequently and regularly, with interest rates determined by the Fed in an attempt to manipulate inflation and recession, followed by immediate reaction by the stock market in response to whichever direction the variable interest takes. In a brochure by Thomas D. Schauf, a certified public accountant, the Fed is exposed totally. He encourages reprinting his brochure freely, so I have quoted some of his information below:

> The Federal Reserve Bank (Fed) can write a check for an unlimited amount of money to buy government bonds and the US Treasury prints the money to back up the check. UNBELIEVABLE... IT'S TRUE. Read *National Geographic*, January 1993, pg. 84. Go to the library and read

the books exposing this SCAM. The Fed is a private bank for profit…just like any business. Check the *Encyclopaedia Britannica* or, easier yet, look in the *1992 Yellow Pages*. The Fed is listed under COMMERCIAL BANKS, not GOVERNMENT. The Fed is no more a *Federal* agency than *Federal* Express.

Schauf goes on to explain some of the intricacies of how the Fed prints currency, then sells it to the government in exchange for government bonds, on which we pay interest, and that become part of the national debt that we, the taxpayers, are obligated to repay. Then the Fed sells these instruments to others, including foreign agencies, spreading around, as it were, our debtors. It seems the Fed doesn't want to keep all its eggs in one basket, or, to put it another way, if it spreads around the debt, it also spreads around the risks. Just remember, as it sells off this debt and receives funds or obligations for funds in exchange, it hasn't yet done anything to earn this money but authorize the printing of the currency. A pretty good scam, isn't it? For a more detailed account of how all this works, I recommend reading Wright Patman's *A Primer on Money*.

This system can be bypassed; we can print our own money, as the Constitution requires. "On June 4, 1963, President Kennedy (JFK) [issued] Executive Order 11110 [one of the rare instances when this executive power was used for something worthwhile] [and] printed *real* US dollars with no debt or interest attached, because he bypassed the Federal Reserve Bank! Upon his death the printing ceased and the currency was withdrawn. Want proof? Ask any coin-dealer for a 1963 Kennedy dollar… It says 'United States

Note,' NOT 'Federal Reserve Note'" (Thomas D. Schauf, CPA, FED-UP™, in a tax petition letter to concerned citizens, viewable here: last accessed September 25, 2013, http://karws.gso.uri.edu/Marsh/Jfk-conspiracy/FEDUP.TXT).

Another major faction involved in world control through economics and trade is the Trilateral Commission (founded 1973, by David Rockefeller), the purpose of which was "to promote world government by encouraging economic interdependence among the [three] superpowers [North America, Japan, and Europe]" ("New World Order Quotes—1975–79," http://www.overlordsofchaos.com/html/new_world_order_quotes_1975-79.html?&lang=en_us). Rockefeller selected Zbigniew Brzezinski (later to become President Jimmy Carter's national security advisor) as the commission's first director and invited President Carter to become a founding member. In his book, *Between Two Ages*, Brzezinski calls for a new international monetary system and a global taxation system. He praised Marxism as a "creative stage in the maturing of man's universal vision" and quoted New Ager de Chardin.

By 1979, just six years after the TLC was founded, its activities were already known well enough to be addressed by retiring Arizona senator Barry Goldwater in his autobiography, *With No Apologies*. Goldwater writes:

> In my view the Trilateral Commission represents a skillful, coordinated effort to seize control and consolidate four centers of power—political, monetary, intellectual, and ecclesiastical. All this is to be done in the interest of creating a more peaceful, more productive world community. What

the Trilateralists truly intend is the creation of a worldwide economic power superior to the political governments of the nation-states involved. They believe the abundant materialism they propose to create will overwhelm existing differences. As managers and creators of the system, they will rule the future.

There are three main democratic and industrialized zones: Europe, Japan, and America (including Canada). These three areas make up the Trilateral Commission, to which approximately 325 important figureheads with varying responsibilities belong.

In 1973, upon the initiation of the first triennium (of the Trilateral Commission), during a period of great tension between world governments, the supposed principal function of this group of distinguished dignitaries was to draw the nations together and examine more closely and cooperatively the difficulties facing each area.

Under the surface, the United States had begun to impress that the American government was losing its sovereignty as international leader. The founders of the TLC saw bringing Europe and Japan in as equals as the obvious solution toward the common goal of a smoothly operating international system, one that would achieve to plot a more successful and mutually beneficial course for the nations in the coming years as challenges arose.

The rise of Japan and the emergence of the European Community (EC) dramatizes the importance of such shared leadership in the eyes of TLC members. And instead of seeing the breakup of the Soviet Union as beneficial, the TLC viewed it as the receding Soviet threat dissolving the "glue" holding the regions together as it began its 1991–94 triennium. They claim that handling economic tensions among our countries and sustaining the advantages of a global economy would be more of a struggle and demand upon those actively involved than it had prior.

From the 325-plus membership, an executive committee is selected, including the chairmen (one from each side of the triangle), deputy chairmen, and thirty-five others. Once each year, the full commission gathers in one of the regions; recent meeting locations have included Paris, Washington, DC, and Tokyo. Members insist these are not secret meetings, and as mentioned earlier, Henry Kissinger was interviewed by the media when entering a hotel conference room at one meeting, where he pointed out that the commission was not a secret organization; its meeting times and locations were in plain view for all to see. Of course, he failed to mention that the media representatives were not welcome at the meeting, nor would information discussed behind closed doors be made available for public discussion.

According to the TLC *Organization and Policy Program* publication, "A substantial portion of each annual meeting is devoted to consideration of draft [task] force reports to the Commission. These reports are generally the joint product of authors from each of the three regions who draw on a range of consultants in the course of their work. Publication follows discussion in the Commission's annual meeting" (Antony Sutton, "Trilaterals Over America," *CPA Book Publishers*, 1995; pdf of book available here: http://www.globalistagenda.org/download/TrilateralsOverAmerica.pdf). Although a number of changes are usually made after discussion, the authors are solely responsible for their final text.

In addition to task force reports, the commission considers other issues in seminars or topical sessions at its meetings. A wide range of subjects has been covered, including the social and political implications of inflation, prospects for peace in the Middle East, macroeconomic policy coordination, nuclear weapons proliferation, China and the international community, employment/unemployment trends and their implications, and the uses of space. Relations with developing countries have been a particular concern of the commission, and speakers from developing countries have addressed each annual meeting since 1980.

Task force reports are distributed only "to interested persons inside and outside government." The same is true for the publication issued on each annual meeting.

The Bilderbergers

"Last, but not least" is a phrase that definitely fits this group. The Bilderbergers are power brokers of the world, and that is no exaggeration.

The Bilderbergers were founded in 1954 by Queen Juliana's husband, Prince Bernhard of The Netherlands. Queen Juliana was "among the first endorsers of 'Planetary Citizens' in the 1970s. Numerous leading Americans have been Bilderbergers, including **Dean Acheson, Christian Herter, Dean Rusk, Robert McNamara, George Ball, Henry Kissinger,** and **Gerald Ford**" (Dennis L. Cuddy, "The New World Order—A Critique and Chronology," *Argumentations*, January 1, 1992; viewable here: http://www.argumentations.com/Argumentations/StoryDetail_7173.aspx.) (Try to observe instances of cross-membership names with the CFR, the TLC, Skull & Bones, etc.)

The Bilderbergers, funded by major one-world institutions, was created to regionalize Europe. The Treaty of Rome, which established the Common Market (today's European Community), was produced at Bilderberger secret meetings. Cross-membership with the CFR includes **David Rockefeller, Winston Lord** (State Department official and Bones family tree), **Henry Kissinger, Zbigniew Brzezinski, Cyrus Vance, Robert McNamara** (former World Bank president), **George Ball** (State Department and director of Lehman Brothers), **Robert Anderson** (ARCO president), **Gerald Ford, Henry Grunwald** (managing editor, *Time*), **Henry J. Heinz II, Theodore Hesburgh** (former Notre Dame president), and others.

Some reliable unnamed sources have provided a pipeline into the secret organization's meetings and furnished copies of the "not for circulation" agenda and roster of attendees to reporters from the *Spotlight*. Again, the names read like a *Who's Who* from around the world. In a report on the Bilderbergers meeting at Baden-Baden, Germany, June 6–9, 1991, reporter James P. Tucker, Jr. writes:

> The Bilderbergers group plans another war within five years.
>
> This grim news came from a "main pipeline"—a high-ranking Bilderberg staffer who secretly cooperated with our investigation—behind the guarded walls of the Badischer Hof, who was operating from inside with colleagues serving as "connecting pipelines."…
>
> While war plans were being outlined in "Bilderbergese," the air traffic controller at Baden-Baden's private airport

reported numerous incoming flights from Brussels, where NATO headquarters are based....

Aboard one of those planes, *en route* to the Bilderberg meeting, was Manfred Woerner, NATO's general secretary.

It was repeatedly stated at the Bilderberg meeting that there will be "other Saddams" in the years ahead who must be dealt with swiftly and efficiently.

What the Bilderberg group intends is a global army at the disposal of the United Nations, which is to become the world government to which all nations will be subservient by the year 2000.

Crucial to making the U.N. a strong world government...is to bestow it with "enforcement powers." "A U.N. army must be able to act immediately, anywhere in the world, without the delays involved in each country making its own decision whether to participate, based on parochial considerations," said Henry Kissinger...[who] expressed pleasure over the conduct of the Persian Gulf war, stressing that it had been sanctioned by the U.N., at the request of President George Bush, himself a Trilateral luminary, before the issue was laid before the U.S. Congress.

The fact that the president would make his case to the U.N. first, when the Constitution empowers only Congress to declare war, was viewed as a significant step in "leading Americans away from nationalism."...

It was "good psychology" for Bush to allow congressional and other leaders to express their fear of losing 20,000 to 40,000 American lives [in the Gulf War]…when Bush knew the loss of life would be much lower.

When the allied casualty toll reached "only 378" and Americans read and heard of "only four" Americans dying in a week of ground war, it "was like nobody had died at all," one said, "and Americans enjoyed it like an international sporting match."

Such an adventure was essential to getting Americans into "the right frame of mind for the years ahead," said another.… They promised each other, there will be "more incidents" for the U.N. to deal with in the years ahead. The Bilderberg group and its little brother, the Trilateral Commission, can set up "incidents" on schedule, they said, but in less direct words. The words "within five years" were heard repeatedly.

Another important step toward a strong, recognized, and accepted world government is taxing power.…

At its April meeting in Tokyo, the Trilateralists called for a U.N. levy of 10 cents per barrel of oil coming from the Persian Gulf. It would be as "temporary," lasting only long enough to rebuild Kuwait and feed the Kurds until they are back on their feet.

The Bilderbergers approved of the move by their brother group, in which Rockefeller and Kissinger…serve as leaders. Once people get used to a tax, it never is repealed…it could be extended worldwide "with appropriate increases" in the years ahead.

From the sum total of all things said, the Bilderberg strategy emerged: Start the tax by imposing it on a newly established "bad guy" who must suffer, and use the revenue for such humanitarian purposes as feeding the Kurds. Keep the initial tax so low that the public is unaware that it is levied. Then kick it up. (James P. Tucker, Jr., "World Shadow Government Planning for Another War," *Spotlight*, September 1991)

Also discussed was the dividing of the world into major regions, eliminating individual countries' borders, "for convenience of administration." Then the group was to apply pressure to the US to pass the free trade treaty with Mexico, another step toward establishing the Western Hemisphere as another region—first free trade with Canada, then Mexico, followed by all other Latin American nations.

The plan was to have a single currency for all of Europe by 1996, with a one-currency movement for the Western Hemisphere to follow, and ultimately a world government with world currency. The Bilderbergers have expressed their pleasure with the progress of the trade agreements/treaties that were underway.

chapter four

Barcodes, GPS, RFID, and the Beastly Biochips of Tomorrow

Barcodes

In and of itself, there is nothing inherently evil about a barcode—it is just another, more efficient way of keeping track of your inventory and movement of freight, laundry, etc.—chores formerly accomplished with painstaking manual labor. The problem lies in placing this technology at the disposal of a New World Order-minded government that not only wants to meddle in our affairs, but wants to control them (and us).

A barcode is simply a mathematically arranged symbol of vertical lines. It is a parallel arrangement of bars and spaces of varying width. The structural arrangement or spacing of these lines, relative to a given set of parameters, can be made to represent a product's identification number. The UPC, better known simply as a "barcode," has been put to this use since 1973, having been adopted by the retail industry in 1972. The UPC barcode is considered the "mother" of all microchip transponder technology.

The code is not really as complicated as it may first seem. A typical UPC *Version A* barcode symbol contains thirty black vertical lines (called "bars"). A *pair* of these lines equals only *one digit* or number. In other words, this coding system *requires two lines to equal one number.*

The UPC code is mainly comprised of ten digits, the first set of five digits linking to the manufacturer of the product, and the second set of five digits linking to distinctive information about the product (or a unique product identifier code). The number to the left preceding the central ten digits is a "number system digit," and the number to the right of the central ten digits is a "transpositional check digit" (or TCD) to flag any transposition errors.

On each side and in the center of this series of carefully spaced lines are three pairs of longer lines that extend slightly below the others. These longer pairs of lines are special—they are called "guard bars."

Guard bars provide reference points for store computer scanners (barcode reading machines) by segregating the left half of the code lines from the right half. This is needed because the left lines have a different message on them than the right lines and must, therefore, be read by the scanner differently. The center pair of guard bars in the UPC barcode is used both to divide the code in half and to tell the scanner/computer what it needs to know in order to readjust its program to interpret the remaining half of the code. Note again the diagram—the left half represents the manufacturer's code and the right half represents the product code.

Simply stated, a typical UPC *Version A* barcode symbol consists of two halves, representing a total of twelve numeric digits. These six-digit halves are surrounded by left, center, and right guard bar patterns. Together, the black and white lines produce a series of electronic "dits" and "dahs," similar to the old Morse code telegraph key "dots and dashes." The scanner reads these electronic messages as a series of "zeroes" and "ones." These zeroes and ones represent what are called *binary numbers,* in computer language.

Binary numbers are a "machine language" that is the basic internal language of computers. This is how the computer "thinks" and gives itself orders. All computers operate on the basis of this two-digit mathematical binary system.

While such barcodes alone are unlikely to be the Mark, they have served to condition mankind for nearly everything in the world—both animate and inanimate—being marked, tagged, coded, numbered, implanted, and identified with some type of ID system that will allow the dictatorial New World Order global government to label, trace, track, monitor, and control everything and everybody on Earth.

Global Positioning Systems (GPS)

This subsection just as well could be called "The Crowded Skies." That's because without all those satellites up there, a Global Positioning System (GPS) would not be possible. But they are, indeed, up there—in abundance. In fact, they seem to be proliferating like little rabbits.

A simplified definition of GPS could be: the ability to locate and track people or things on a global scale, to know the exact position of anything, utilizing a battery-powered GPS receiver or other similar telecommunications device (or, as discussed earlier, almost any electronic gadget in any purse or back pocket of the average American).

How does GPS operate, and why is it needed? The answer to the first part of that question is very technical, and the answer to the second part is very obvious. Simply stated, signals are sent from a series of satellites from space to Earth. Receivers on Earth triangulate the signals and calculate latitude/longitude position. The information may then be fed into a computer controlled by the individual (or organization) doing the "locating."

The possibilities for commercial use of this technology appear to be, and have already proven to be, limitless…which means the possibilities for abuse of this technology likewise appear to be limited only to the imagination of the powers who control it. With Big Brother gaining more control and power every day, what do *you* think will be the eventual use of this kind of technology?

Winn Schwartau, in his book, *Information Wars*, states:

The question "Where are you?" will be answered at the push of a button. Global positioning satellites will know, to within a few feet, your exact location. Lives will be saved as personal digital assistants broadcast the location of lost or injured or kidnapped people. But what about employees?

Will their every step be tracked to enhance security or to evaluate their performances for promotions? To the dismay of the unions who say the practice is an invasion of privacy, we already track the routes and times of trucks to increase shipping efficiency. Computers already know almost everything about us; will we also decide to add our every location to this list?

The May 8, 1994, edition of *The Bulletin*, Bend, Oregon, carried an article by writer Ralph Vartabedian entitled "Defense Satellite Technology Ready for Commercial Boom." According to Vartabedian, "The Pentagon is awash in obsolete nuclear bombs, mothballed battleships, and surplus military bases [although FEMA apparently has plans to make use of these deserted bases], but out of the scrap heaps left by the Cold War has come a technology with a promising payoff." Below are some excerpts.

When the Defense Department laid plans in the 1970s for its Global Position System, a network of 24 satellites that broadcasts navigation signals to users on Earth, it was intended to help soldiers fight anywhere, from jungles to deserts.

Along the way, though, commercial interests saw a potentially lucrative concept that could revolutionize industries such as land surveying, trucking, environmental protection, and farming.

The technology [is] now poised to leap into virtually every facet of the American economy....

With a special receiver that taps the satellite signals, civilian users can determine their position by latitude and longitude within 100 meters (328 feet) anywhere in the world. Once as big as a file cabinet, the receivers are now the size of a paperback book and still shrinking....

Some visionaries anticipate the day when virtually everything that moves in U.S. society—every shipping container, aircraft, car, truck, train, bus, farm tractor, and bulldozer—will contain a microchip that will track and, in many cases report its location. [Note: This is no longer just the dream of some visionaries—it's actually occurring now, and Vartabedian forgot to mention they'll be tracking your garbage, as well.] Massive computer systems, they say, will tie together the movement of assets in the economy, providing a sophisticated information system for the status and location of goods.

"Communications satellites were the first great success in space, but GPS is going to dwarf that," said...a former Hughes Aircraft chairman.... "GPS is going to pervade everything we do."...

Eventually the price will drop below $50.... At that point, GPS would be inserted into a lot of other electronic gear [sound familiar?]...that could instantly alert police [or others] to an individual's location....

Computerized maps are being used to track the spread of disease, pollution, and crime, based on data collected from GPS systems. Hamburger chains pour over these kinds of computer-generated maps to determine the best sites for new franchises.

Interstate truckers use the system to keep tabs of their road taxes.... Cities use the satellite system to dispatch emergency vehicles and track the location of passenger buses. Railroads are finally able to figure out where their trains are.

Orbiting 11,000 miles above Earth, the 24 satellites are the heart of the system....

An article appearing in ENR entitled "Surveying's Brave New Digital World" included the following preface:

Surveying and mapping tools have progressed significantly from the days of meticulously entering transit readings in survey notebooks. From the latest in survey marker technology to the latest in computer-enhanced technologies, surveyors and engineers can do their jobs faster, better, and with more precision than ever before.

Recent hardware advances, declining prices of hardware and software, and the greater availability of pre-packaged data are making geographic information system (GIS) technology an affordable and appealing technology for even the smallest firms. In some instances, state legislatures are funding new initiatives or enacting legislation requiring GIS use throughout the state, and Public Utility Commissions are mandating that utilities use GIS to ensure efficient and low-cost public services. Consultants hoping to contract services to these organizations will have to move into the "all digital" world or risk obsolescence. Many private sector clients—such as large engineering firms or developers—are also requesting surveying firms use computer-aided design (CAD) or GIS on projects and deliver digital products.

In this year's special section on surveying and mapping, we'll look at what's new today and what lies just ahead in products and services—including some exciting trends for users of GIS technology, *global positioning systems* (GPS), and satellite imagery/aerial photography. (emphasis added)

The application of GPS technology can run from the mundane to the "Indiana Jones" adventure project. For example, one company was contracted to map a Caribbean island about ten thousand feet off the coast of St. Thomas. Since it was used for jungle warfare training during and following World War II, there was a real possibility of unexploded ordnance remaining behind. "Lowe [the engineering firm] used the latest in both hazardous materials handling methods and surveying—including GPS, aerial photography, and GIS technology—to produce preliminary digital (CAD) maps in just 30 days." Gone are the "good old days" when two guys stood behind tripods and waved at each other... for the most part.

One of the major uses of all those satellites is telecommunications, in a myriad of forms. In *Infomart* magazine, AT&T's Roy Weber, director of new business concepts, asks the question, "What's This World Coming To?" I'd be happy to tell him, but since he's the director for new business concepts, I'm sure he's one of those visionaries who is limited only by his imagination when it comes to what we can do with our technology in the future.

Weber has promoted a six-theme program running the gamut from visual communications and telephone automated voice recognition to "connecting the world" with a multimedia broadband network and numbering *you*. Of course, that's my favorite. It is Weber's Theme Five: "We do everything wrong. We number telephones. You don't want to speak to the telephone. You want to speak to a person…we're developing Global Personal Calling Services that will find you [anywhere]."

The editor actually felt obliged—I'm sure because of the content of the article—to insert a somewhat tongue-in-cheek preface between the title and the start of the article. It begins:

> **Editor's note:** In the harsh light of the hospital nursery, the nurses are filling in the blanks on a birth certificate: name, weight, phone number. Phone number? The mere thought of assigning a number to a human being calls forth torrents of ethical and philosophical questions—questions about the relationship between man and machine, about individuality, about privacy.

For Weber, these questions are not simply a matter of intellectual curiosity, they are part of the exercise of inventing the future at AT&T.…

And may we remind you that AT&T has always been right up there with the big banks, MasterCard, and Visa when it comes to development and promotion of smart cards.

Let's see what Schwartau had to say on the subject of satellite telecommunications:

In under thirty years, satellite communications became an absolute necessity for international transactions. Today, the demand is such that hundreds of new satellite launches are being planned. Motorola's Iridium Project, for example, will ring the planet with sixty-six satellites, permitting portable phone users to talk to anyone, anywhere, at any time. [Can you imagine how amazing and sensational this announcement was at the time? How desensitized technology has made us all?] A true multinational effort is under way, including Japanese money and manufacturing and Russian orbital launch capabilities. Two competing consortiums have also begun staging their own satellite-based competitive global communications efforts. ...

Charles Reich also noticed that technology and society were at odds. "What we have is technology, organization, and administration out of control, running for their own sake. ... And we have turned over to this system the control and direction of everything—the natural environment, our minds, *our lives*." (emphasis added)

It's hard to say it much better than that. But as the title of Schwartau's book, *Information Warfare*, implies, he continues to issue warnings about the pitfalls involved:

Not all of the switch connections are made through and across wires of copper and fiber optics. Communications increasingly uses the airwaves, as we can see in the proliferation of cellular phones, Motorola's multibillion dollar

Iridium Project, and microwave and satellite transmissions. The electromagnetic ether represents a new battlefield for the Information Warrior.

Cellular phone conversations, for example, are wide open to interception by $179 scanner devices that can be bought from Radio Shack, *Monitoring Times* magazine, or dozens of other sources. Courts have upheld that there is no reasonable expectation of privacy when one is talking on a cellular phone.

Then Schwartau proceeds to describe a number of different methods used in *telefraud*, and to tell how much it costs the public. If you are convinced that your secrets (or any other information) are safe—even with encryption devices, because the government has the "back-door key"—on the Internet, the information super-highway, satellite telecommunications, your smart card, the Social Security computers…you're really just kidding yourself—probably because it's too scary to consider the alternative.

Radio-Frequency Identification (RFID)

Most readers are undoubtedly familiar with the development of radio-frequency identification (RFID) technology that, under certain applications, is forecast to be connected to future human-enhance-ment technologies, especially neurosciences, brain-machine inter-facing, and cybernetics.

These RFID chips employ tiny integrated circuits for storing and processing information using an antenna for receiving and transmitting the related data. This technology is most commonly applied as a "tag" for tracking inventory with radio waves at companies like Walmart, where consumer goods are embedded with "smart tags" that are read by handheld scanners for supply-chain management.

In recent years, RFID technology has been expanding within public and private firms as a method for verifying and tracking people as well. We first became aware of this trend a while back when a chief of police—Jack Schmidig of Bergen County, New Jersey, a member of the police force for more than thirty years—received a VeriChip (RFID chip) implant as part of Applied Digital Solution's strategy of enlisting key regional leaders to accelerate adoption of its product.

Kevin H. McLaughlin (chief executive officer of VeriChip Corp. at the time) said of the event that "high-profile regional leaders are accepting the VeriChip, representing an excellent example of our approach to gaining adoption of the technology" (note that VeriChip Corp. was renamed to PositiveID Corp. on November 10, 2009, through the merger of VeriChip Corp. and Steel Vault Corp.). Through a new and aggressive indoctrination program called "Thought and Opinion Leaders to Play Key Role in Adoption of VeriChip," the company set out to create exponential adoption of its FDA-cleared, human-implantable RFID tag. According to information released by the company, the implantable transceiver "sends and receives data and can be continuously tracked by GPS (Global Positioning Satellite) technology." The transceiver's power supply and actuation system are unlike anything ever created. When implanted within a body, the device is powered electromechanically through the movement of muscles and can be activated either by the "wearer" or by the monitoring facility. In the wake of the terrorist attacks in New York City and Washington, DC, an information technology report highlighted the company's additional plans to study implantable chips as a method of tracking terrorists. "We've changed our thinking since September 11 [2001]," a company spokesman said. "Now there's more of a need to monitor evil activities" ("Will You Grin for the RIFD Mark of the Beast?" *Before It's News*, October 26, 2010, http://beforeitsnews. com/alternative/2010/10/will-you-grin-for-the-rfid-mark-of-the-beast-236019.html). As a result, PositiveID has been offering the company's current incarnation of implantable RFID as "a tamper-proof means of identification for enhanced e-business security… tracking, locating lost or missing individuals, tracking the location of valuable property [this includes humans], and monitoring the medical conditions of at-risk patients." While PositiveID offers testimony that safeguards have been implemented to ensure privacy in connection with its implantable microchips, some believe privacy is the last thing internal radio transmitters will protect—that, in fact, the plan to microchip humanity smacks of the biblical Mark of the Beast. Has an end-times spirit indeed been pushing for adoption of this technology this generation?

Consider the following:

- According to some Bible scholars, a biblical generation is forty years. This is interesting, given what we documented in our book, *Zenith 2016: Did Something Begin in the Year 2012 that Will Reach Its Apex in 2016?*, concerning the Jewish Calendar year 5773 (2012—but 2013 in the most commonly used Gregorian calendar), from which, counting backward forty years, one arrives at the year 1973, the very year *Senior Scholastics* began introducing school kids to the idea of buying and selling in the future using numbers inserted in their foreheads. In the September 20, 1973, feature "Who Is Watching You?" the secular high school journal speculated:

All buying and selling in the program will be done by computer. No currency, no change, no checks. In the program, people would receive a number that had been assigned them tattooed in their wrist or forehead. The number is put on by laser beam and cannot be felt. The number in the body is not seen with the naked eye and is as permanent as your fingerprints. All items of consumer goods will be marked with a computer mark. The computer outlet in the store which picks up the number on the items at the checkstand will also pick up the number in the person's body and auto-matically total the price and deduct the amount from the person's "Special Drawing Rights" account. (See: "Storm Clouds," *WattPad*, 2013: http://www.wattpad.com/147636-storm-clouds?p=86.)

- The following year, the 1974 article, "The Specter of Eugenics," had Charles Frankel documenting Nobel Prize-winner Linus Pauling's suggestions that a mark be tattooed on the foot or forehead of every young person. Pauling envisioned a mark denoting genotype.

- In 1980, *US News and World Report* revealed that the federal government was plotting "National Identity Cards" without which no one could work or conduct business.
- The *Denver Post Sun* followed up in 1981, claiming that chip implants would replace the identification cards. The June 21, 1981, story read in part, "The chip is placed in a needle which is affixed to a simple syringe containing an anti-bacterial solution. The needle is capped and ready to forever identify something—or somebody" ("Will You Grin for the RFID Mark of the Beast?").
- The May 7, 1996, *Chicago Tribune* questioned the technology, wondering aloud if we would be able to trust "Big Brother under our skin?"
- Then, in 1997, applications for patents of subcutaneous implant devices for "a person or an animal" were filed.
- In August 1998, the BBC covered the first-known human microchip implantation.
- That same month, the Sunday Portland *Oregonian* warned that proposed medical identifiers might erode privacy rights by tracking individuals through alphanumeric health-identifier technologies. The startling *Oregonian* feature depicted humans with barcodes in their foreheads.
- Millions of *Today Show* viewers then watched in 2002 when an American family got "chipped" with Applied

Digital Solution's VeriChip live from a doctor's office in
Boca Raton, Florida.

- In November of the same year, IBM's patent application
  for "identification and tracking of persons using RFID-
  tagged items" was recorded.

- Three years later, former secretary of the Health and
  Human Services department, Tommy Thompson,
  forged a lucrative partnership with VeriChip Corp. and
  began encouraging Americans "to get chipped" so that
  their medical records would be "inside them" in case of
  emergencies.

- The state of Wisconsin—where Thompson was
  governor before coming to Washington—promptly drew
  a line in the sand, passing a law prohibiting employers
  from mandating that their employees get "chipped."
  Other states since have passed or are considering
  similar legislation. Despite this, in the last decade, an
  expanding number of companies and government
  agencies has started requiring the use of RFID for people
  identification. Unity Infraprojects, for example, one of
  the largest civil contractors in India, tracks its employees
  with RFID, as does the US Department of Homeland
  Security for workers involved in baggage handling at
  airports.

- Since September 11, 2001, the US government has
  proposed several versions of a national ID card that
  would use RFID technology.

- Since 2007, the US government began requiring all

passports to include RFID chips that enable use of biometric features such as facial recognition.

- Hundreds of Alzheimer's patients have been injected with implantable versions of RFID tags in recent years.

- RFID bracelets are now being placed on newborns at a growing list of hospitals.

- Students are being required in some schools and universities to use biometric ID employing RFID for electronic monitoring.

- Thousands of celebrities and government officials around the world have had RFID radio chips implanted in them so that they can be identified—either for entry at secure sites or for identification if they are kidnapped or killed.

- Others, like Professor Kevin Warwick (a British scientist and professor of cybernetics at the University of Reading in the United Kingdom), have been microchipped for purposes of controlling keypads and external devices with the wave of a hand.

- Besides providing internal storage for individual-specific information like health records, banking and industry envisions a cashless society in the near future where all buying and selling could transpire using a version of the subdermal chips and wireless authentication. As mentioned above, in 1973, *Senior Scholastics* magazine introduced school-age children to the concept of buying and selling using numbers inserted in their foreheads. More recently, *Time* magazine, in its feature story, "The

Big Bank Theory and What it Says about the Future of Money," recognized that this type of banking and currency exchange would not require a laser tattoo. Rather, the writer said, "Your daughter can store the money any way she wants—on her laptop, on a debit card, even (in the not-too-distant future) on a chip implanted under her skin" (Joshua Ramo, "The Big Bank Theory," *Time*, April 27, 1998; see: http://www.time.com/time/printout/0,8816,988228,00.html).

- In 2007, PositiveID, which owns the Food and Drug Administration-approved VeriChip that electronically transmits patients' health information whenever a scanner is passed over the body, ominously launched "Xmark" as its corporate identity for implantable healthcare products.

- Skip forward to the present, and suddenly the push for a national biometric identification system and RFID technology is all over the news and within industry trade reports. The *Next Generation Biometric Market—Global Forecast & Analysis 2012–2017* noted that the global biometric identification market was surging toward the nearly $14 billion mark by 2017, with an estimated compound annual growth rate of 18.7 percent.

- In February 2013, a report for the Competitive Enterprise Institute authored by David Bier ("The New National Identification System Is Coming") documented how US lawmakers including Senator John McCain and Senator Lindsay Graham were advocating for a "super" identification system that would include biometrics.

- Three months later, in May, the Massachusetts-based engineering firm MC10 disclosed that it is developing a high-tech, biostamp, electronic "tattoo" that will replace all passwords. It is made of silicon and is sealed on the wearer's body. As this book heads to the editor, MC10 hopes to have its first prototypes "affixed" to humans with the next few months.

- Simultaneously, Motorola Senior Vice President Regina Dugan announced that a project similar to MC10's is now under development at the multinational telecommunications company. Called "The Proteus Digital Pill," the project contains a computer chip and transmits an 18-bit, ECG-like signal that can communicate with mobile devices as well as serve as a biometric ID. The ingestible "pill" has already been approved for human use and tracking by the FDA in the United States as well as in Europe. Note that "Proteus" is a shape-shifter and primordial pagan god of ancient sages (seers) that can affect the "conscious" *and is capable of mutating the host.*

- No sooner had Motorola announced its plan for the "Proteus" swallowable marker than some in the secular media marched forward to brand any concerned or resistant religious types (such as the authors of this book) as inflexible shrills who do not represent true Christianity. As an example, Iain Thomson of *The Register* wrote on May 31, 2013:

One marketing problem Motorola may not have anticipated is the reaction of biblical literalists to its…authentication systems.

A surprising number of people in the US still adhere to an apparent literal translation of the current version of… the finale of the New Testament: The Book of Revelation—or, for you believers of the Catholic persuasion, The Apocalypse.

The text, thought to be written about 60 years after the biblical death of Christ, is regarded as either a description of the end times of humanity, a satirical pastiche on the increasingly subverted tenets of Christian bureaucracy, or a really bad mushroom trip on a Greek island. Nevertheless it contains the following warning:

> "And he causeth all, both small and great, rich and poor, free and bond, to receive a mark in their right hand, or in their foreheads: And that no man might buy or sell, save he that had the mark, or the name of the beast, or the number of his name."

Be reassured that the majority of people of faith in the US and elsewhere aren't quite so inflexible. Those that aren't may be shrill, particularly in the US, but do not form a representative sample of Christianity.

- In January, 2013, Robert A. Pastor, professor of international relations at American University, argued before the US Congress that the majority of Americans are now ready for—and need—a national ID based on head and hand biometrics. A centerpiece of the immigration bill imagines just such a scenario and would require all citizens to have a biometric card, without which no one would be approved for employment (effectively rendering him or her as a "non-person").

- By June, 2013, the Congress of the United States pushed forward on related mandates, demanding progress on advanced biometric smart cards for federal identification under the Homeland Security Presidential Directive (HSPD-12). The DHS wants these personal verification IDs immediately and for its employees to carry digitalized finger (and/or palm) and facial recognition images (head and hand) to serve as the trendsetter for all levels of government and private industry identification. These cards will employ barcodes, RFID tags, and onboard data processors that can transmit information and location to remote sites.

- Later the same month, the use of biometrics (hand and/or head-iris scanning) as a payment option for goods and services was documented as the method of choice for buying and selling among 50 percent of consumers, with that percentage trending upward (see Revelation 13:17).

- One month later, in July, a special report conducted by Natural Security (a UK-based authority in user-authentication) described nine hundred consumers who had participated in a pilot program in which they used fingerprint-based technology when purchasing products and services. Of that number, 94 percent of them were excited about the scheme, agreeing they are now ready to use biometrics and RFID technology for all buying and selling.

- By August of 2013, a new report from the National Institute of Standards and Technology confirmed the long-term viability of iris recognition as stable for biometric identification and that no distinguishing texture or degradation of the iris occurs for at least ten years, if not decades.

- A whole host of personal products began flooding the market at the same time, including jewelry, headgear, and glasses that boast GPS and RFID tracking capabilities, with promises of future payment integration for buying and selling via biometric signatures. One example is the new "Nymi Bracelet" that is worn around the wrist. It monitors the heartbeat as a biometric signature, claims to be more accurate than facial recognition, and is said to be about as accurate as fingerprint scanning.

- Also at the time this book is heading to the printer, nineteen US states have complied with the "Real ID Act," an act of Congress that modifies US federal laws pertaining to authentication standards for driver's licenses and identification cards with the goal of codifying a national—then international—biometric ID system.

- Concurrently making its way through Congress as part of an immigration reform bill is a provision to mandate universal biometric identification in the form of a national ID, without which nobody will be federally approved for employment (or, as it says in the book of Revelation 13:17, "that no man might buy or sell, save he that had the mark…"). It is called "E-Verify," and incredibly not only has bipartisan support among lawmakers but enthusiastic approval from notable Christian leaders (funded by George Soros, no less). This includes:

  - Leith Anderson, President, National Association of Evangelicals
  - Stephan Bauman, President and CEO, World Relief
  - Jim Daly, President, Focus on the Family
  - Noel Castellanos, CEO, Christian Community Development Association
  - Luis Cortes, President, Esperanza
  - Dr. Richard Land, President, Ethics and Religious Liberty Commission of the Southern Baptist Convention
  - Samuel Rodriguez, President, National Hispanic Christian Leadership Conference
  - Dr. Carl Ruby, Vice President for Student Life, Cedarville University
  - Gabriel Salguero, President, National Latino Evangelical Coalition

> ➢ Mat Staver, Founder and Chairman, Liberty Counsel
> ➢ Jim Wallis, President and CEO, Sojourners
> ➢ The Catholic Church, which, in September 2013, announced a massive, coordinated effort to get the immigration reform bill passed by targeting sixty Catholic House Republicans and one hundred and thirty members of the House who are also Catholic
> ➢ In chorus, other evangelical pastors nationwide broadcast radio ads in fifty-six congressional districts in fourteen states at a cost of $400,000 in support of the plan.

The list above continues to exponentially increase, causing a growing number to wonder if a national ID system including RFID adoption will, for all practical purposes, result in every man, woman, boy, and girl in the developed world having an ID chip inside him or her (like animals worldwide already do) sometime soon. Makers of implantable microchips like to state that the process is voluntary, but a report by Elaine M. Ramish for the Franklin Pierce Law Center says:

A [mandatory] national identification system via microchip implants could be achieved in two stages: Upon introduction as a voluntary system, the microchip implantation will appear to be palatable. After there is a familiarity with the procedure and a knowledge of its benefits, implantation would be mandatory.

George Getz, the communications director for the Libertarian Party at the time, agreed, saying:

After all, the government has never forced anyone to have a [driver's] license, [but] try getting along without one, when everyone from your local banker to the car rental man to the hotel operator to the grocery store requires one in order for you to take advantage of their services, that amounts to a de facto mandate. If the government can force you to surrender your fingerprints to get a driver's license, why can't it force you to get a computer chip implant?

Students of eschatology (the study of end-times events) find it increasingly difficult to dismiss how this all looks and feels like movement toward fulfilling Revelation 13:16–17: "And he causeth all, both small and great, rich and poor, free and bond, to receive a mark in their right hand, or in their foreheads: And that no man might buy or sell, save he that had the mark, or the name of the beast, or the number of his name."

As newer versions of RFID-like transmitters become even more sophisticated—adding other "prophetic" components such as merging human biological matter with transistors to create living, implantable machines—the authors of this book believe the possibility that the Mark of the Beast could arrive through a version of smart-chip technology increases. That is one reason we found the recent *Discovery News* report, "Part-Human, Part Machine Transistor Devised," particularly disturbing:

> Man and machine can now be linked more intimately than ever. Scientists have embedded a nano-sized transistor inside a cell-like membrane and powered it using the cell's own fuel. The research could lead to new types of man-machine interactions where embedded devices could relay information about the inner workings of proteins inside the cell membrane, and eventually lead to new ways to *read, and even influence, brain or nerve cells.*
>
> "This device is as close to the seamless marriage of biological and electronic structures as anything else that people did before," said Aleksandr Noy, a scientist at the University of California, Merced, who is a co-author on the recent ACS Nano Letters. "We can take proteins, real biological machines, and make them part of a working microelectronic circuit." (emphasis added; Eric Bland, "Part-Human, Part-Machine Transistor Devised," *Discovery News*, June 2, 2010, http://news.discovery.com/tech/transistor-cell-membrane-machine.html)

A similar story ("DNA Logic Gates Herald Injectable Computers") was published by *New Scientist* magazine the same month as the story above, and a few weeks earlier, an article by the *Daily Mail* ("Meet the Nano-Spiders: The DNA Robots that Could One Day

Be Walking through Your Body") reported the creation by scientists of microscopic robots made of DNA molecules that can walk, turn, and even create tiny products of their own on a nano-scale assembly line.

chapter five

Resistance Is Futile: The Plan to BORG Humanity

F ans of the television series *Star Trek* will equate such biological tinkering described in the previous chapter with The Borg ("Cyborg"): the greatest villains ever introduced to television audiences. The biological and technological terrors made their debut on May 8, 1989, in the "Q, Who?" episode of *Star Trek: The Next Generation*.

"This is the Borg Collective," they said menacingly. "Prepare to be assimilated. We will add your biological and technological distinctive to our own. You will adapt to service us. Resistance is futile."

Viewers sat on edge as the cybernetically enhanced and immensely powerful humanoids overcame the *USS Enterprise* and her crew. Implanted with biometric devices connected to a sophisticated communications network known as "The Borg Collective," the superior beings moved without conscience to assimilate the Star

Trek crew and implant them with biometric devices designed to facilitate the needs of the Collective.

When the crew of the Enterprise finally escaped during the two-part cliffhanger, Trekkies around the world exhaled a Borg-like collective sigh of relief.

"I AM LOCUTUS OF BORG!"

At one point during *The Next Generation* series, Captain Jean-Luc Picard was captured again and assimilated by the Borg. He became known as Locutus of Borg and promised to "raise the quality of life for all species." This would be accomplished by forced integration into the Collective.

Lately, real-time technology companies have illustrated how life in the not-too-distant future may imitate the filmmaker's art. Applied Digital Solutions (ADS) and other tech firms have international agreements to distribute Borg-like technology in the form of miniature digital transceivers designed for human implantation. Research teams funded by the National Institutes of Health, NASA, and a barrage of privately funded laboratories are also developing implants as well as external neural readers that will make it possible for people to communicate through computers using the power of thought. Tests have recently illustrated this as a viable concept and, in August of 2013, University of Washington researchers even performed the first noninvasive human-to-human brain interface, in which one researcher was able to control the hand of another test subject by sending a brain signal via the Internet to the second researcher. This emergent technology could ultimately facilitate a wide array of complicated tasks assigned by the New Collective.

If this all sounds a bit Orwellian—it is. It is also reality, and many Christians believe such technology points to an Antichrist system that will ultimately assimilate ethnic groups, ideologies, religions, and economics from around the world into a New World Order "Collective." But control of the NOW's "assimilated" will be derived at the expense of individual human liberties. Everyone, "both small and great, rich and poor, free and bond [will be forced to] receive a mark [*charagma*; from Greek *charax* meaning to "stake down into" or "stick into," such as with a hypodermic needle injecting something under the skin!] in their right hand, or in their foreheads: And that no man might buy or sell, save he that had the mark, or the name of the beast, or the number of his name. Here is wisdom. Let him that hath understanding count the number of the beast: for it is the number of a man; and his number is Six hundred threescore and six" (Revelation 13:16–18).

According to experts, experiments in behavior modification have also been conducted using implanted chips and may explain the ruthless methods of assimilation that will be employed by the Borg-like followers of the Antichrist. (See Revelation 13:15; 20:4)

Be Assimilated, or Be Stereotyped and Destroyed!

One cannot read the books of Revelation and Daniel without observing the unique combination of political (Antichrist) and religious (False Prophet) personalities operating within the Antichrist's "collective." How will religious leaders and laypeople be convinced to follow anti-Christian anti-democracy? Enter a pattern reminiscent of Jewish persecution in postwar Germany: isolation of and discrimination against conservatives who fear the loss of individual human liberties. The need to trivialize those who appreciate individual freedoms is necessary, leading to mandatory assimilation.

In Russia, when national tax ID numbers were being introduced, Andrei Zolotov, Jr. wrote in *The Moscow Times*: "Some right-wing Christians fear the growing computerization of the world is opening the way for the coming of the Antichrist. The government's new, widely publicized plan to give every citizen a tax identification number and talk of introducing social security cards with barcodes—dreaded by those who see Satan's number, 666, in the codes—has apparently given them cause for further alarm."

The Holy Synod of the Russian Orthodox Church addressed the government's plans, saying, "Many Christians, who consider the name given to them in baptism holy, consider it unworthy to ask the government for some new 'name' in the form of a number."

But Sergei Chapnin, editor of the Russian magazine *Sobornost Orthodox* (www.sobor.ru), said the religious community's response is a sign that an "occult" mentality is penetrating the Orthodox Church. "To believe in the magic of numbers is absolutely a non-Christian attitude toward life," Chapnin said. "If some people are afraid of it, it only says that occultism is intruding into Christian consciousness, and first of all the consciousness of neophytes who are the majority in today's Russian Orthodox Church."

How convenient.

Locutus of Borg: "We will raise the quality of life for all species."

Work on microchip technology as a method of raising the quality of life through medical advancements is being conducted at laboratories like the Fetal Treatment Center (FTC) at UC San Francisco, where scientists have already successfully connected (NASA's) implantable biotelemetry devices to unborn babies.

Other specialists, such as Dr. Roy Bakay of Emory University in Atlanta, Georgia, are installing chip-to-brain implants.

Charles Ostman, a senior fellow at the Institute for Global Futures, says, "Neuroprosthetics are…inevitable. Biochip implants may become part of a rote medical procedure. After that, interface with outside systems is a logical next step."

Professor Kevin Warwick, the first known recipient of a biometric chip implant, speaks excitedly of microchip implantation. "Right-o, got the signal, got the implant; all I've got to do is run a wire from the implant to my nervous system. I'm so excited about it, I want to get on with the next step straight away. Let's see if we can control computers directly from our nervous system."

When asked about the Borg-like ramifications of such technology, Professor Warwick responds, "It is possible for machines to become more intelligent than humans in the reasonably near future. Machines will then become the dominant life form on Earth."

Spoken like a good little Borg wannabe.

The Role of Transhumanism: Merging Man with the (666) Machine

No discussion of the coming Mark of the New World Order is complete without addressing the biological and technological implications. Now, many of you may prefer to "chapter jump," selecting only those topics that most interest you—and it's a safe bet that most of us spent the majority of our high school science classes daydreaming about extracurricular activities—but I ask you to bear with me during this seminal section.

Science and technology have formed an unholy alliance during recent decades, giving birth to an emerging discipline called biotechnology. It is of paramount importance that every believer knows at least the basics, so consider this your biotech primer. Yes, the topic is complicated, so let's see if we can cut through the Gordian knot of biotechnology. Just remember, the high degree of complexity provides a gateway that few wish to open, which provides cover for an insidious agenda. Trust me; you'll want to read this.

To begin, let's spend a few paragraphs covering the bare bones of biological and technological histories. The etymology of the term "science" is often cited as originating with the Latin noun *scientia*, which means "knowledge." You might even recognize it from the oft-quoted phrase *scientia est potentia*, translated as "knowledge is power." We can find this disturbing little motto blatantly emblazoned upon the logo for the Information Awareness Office (IAO). Early Illuminist Sir Francis Bacon is said to have coined this phrase, but it is more probably the invention of Enlightenment philosopher and amateur physicist, Thomas Hobbes, who included it in the context of his book *De Homine*. It is perhaps not coincidental that Hobbes also wrote a treatise titled *Leviathan*, in which he argues in favor of absolute power of governments to decide the fates of the governed. Of course, the quest for power did not commence in the 1600s, for it was knowledge—or perhaps better said, *secret* knowledge—that the serpent used to tempt Adam and Eve into disobeying God in the first place!

Despite this fact, it can be said that *scientia* also has a decidedly Christian "flavor." The noun *scientia* is derived from the Latin verb *scire*, which means to "separate one thing from another." *Scire* is related to another verb, *scindere*, which means "to divide." As a Bible scholar, you must already be making the connection, for it is a verse many of us have memorized!—2 Titus 2:15, which states, *"Study to show thyself approved unto God, a workman that needs not to be ashamed, rightly dividing the word of truth."* The original language here uses the Greek verb *orthotomeo*, a compound word that combines the words *orthos* and *tomos*. *Orthos* is an adjective meaning "straight, not crooked." *Tomos* is another adjective that means "sharp," but it is based on the verb *temno*, which means "to cut or slice in a single stroke." English speakers use these terms all the time, but we often do not make the connection. Orthodontics refers to straightening teeth. And have you or a loved one ever had a CT scan? The letters stand for Computerized Tomography, which means taking images of you in "slices" using a computer. Both Latin and Greek words paint a picture of someone who seeks to discern truth by slicing to the heart of the matter through diligent study.

This is inherently what God-centered *science* is all about! Did you know that you are practicing "science" of a sort whenever you diligently seek God's truth?

Scientific study and methods of observation and careful documentation are often accredited to the Greeks, but the Mesopotamian peoples of ancient Sumer (modern Iraq) demonstrated a keen understanding of Pythagoras' Law ($a^2 + b^2 = c^2$) as early as the eighteenth century BC. (See the Wikipedia article, "Plimpton 322," covering the history and significance of the cuneiform tablet known as Plimpton 322, discovered sometime before 1922, when archaeologist Edgar J. Banks sold the 2 cm-thick tablet to New York publisher, George Arthur Plimpton: http://en.wikipedia.org/wiki/Plimpton_322; accessed May 22, 2013.) Astronomy is often considered the most ancient science, along with mathematics. Babylonian astrologers tracked the movement of stars and observed the heavens, and the prophet Daniel trained among the Chaldean astrologers and scientists as a well-favored Hebrew captive taken by King Nebuchadnezzar during his invasion of Judah.

Solomon, we're told, made keen observations about nature in his quest to know more about God Almighty. China, India, Arabia—all have long histories that record movements of the stars, patterns among animals and plants, and the intricate and predictable interactions of numbers. Science has been around for a long, long time. However, the concerted study of biological processes is a relatively young field. Medicine is as old as Genesis, but even rudimentary observation of cellular organelles and structure is but a few centuries old. Molecular science is even younger, and genetics younger still. Gregor Mendel recorded patterns of inheritance in pea plants in the mid-nineteenth century. In that same century, Charles Darwin theorized that simple organisms gave rise to more complex ones (an erroneous theory based on observation and assumption). In 1838, Schlieden and Schwann promoted their belief that all life is based on cells. By 1860, most biologists agreed, and the subdiscipline of cytology emerged. Sir Francis Galton (Darwin's cousin) ran with his cousin's theories about inheritance, biometrics, and social class and formed the basis for eugenics.

As the twentieth century dawned, eugenics and the race to identify the cellular source for inheritance began. With the discovery of Mendel's research, eugenics as a science shaped politics and public conscience. The prophet Daniel's vision (chapter 12) of the final years before Christ's return came with a caveat: "But you, Daniel, shut up the words and seal the book, until the time of the end. Many shall run to and fro, and knowledge shall increase" (Daniel 12:4). The twentieth century certainly fit this definition. Consider these landmark events (the list is lengthy, but take the time to read it through):

- 1900: The first successful radio receiver transmission
- 1902: The lie detector; neon lights
- 1903: The Wright Brothers' gas-powered airplane engine
- 1904: Vacuum diode; tractor
- 1905: Albert Einstein publishes theory of relativity
- 1906: Lewis Nixon invents first sonar device; Lee de Forest invents triode
- 1907: First helicopter; synthetic plastic
- 1908: Ford's Model T (assembly line); Geiger counter; gyrocompass

- 1910: Edison invents talking movies
- 1912: Military tank invented
- 1913: Merck invents drug now known as "ecstacy"; Bertrand Russell writes *Principia Mathematica*, founding a new artificial intelligence quest
- 1914: Morgan gas mask just in time for World War I
- 1915: Chess automaton built
- 1916: Stainless steel; radio tuners
- 1918: Superheterodyne radio circuit invented (used even today in TV and radio); Spanish flu pandemic
- 1919: Short-wave radio
- 1920: The tommy gun
- 1921: Diphtheria vaccine; tuberculosis vaccine
- 1923: Television; frozen food; term "robot" coined in the play "Rossum's Universal Robots"
- 1924: Scarlet fever vaccine
- 1926: Liquid-fueled rockets; pertussis vaccine
- 1927: Technicolor; iron lung
- 1928: Penicillin; discovery that bacteria can transfer genetic information
- 1930: Jet engine
- 1931: Electron microscope; radio telescope; chromosome linkage discovered that confirms chromosomes carry heritable information
- 1932: Yellow fever vaccine
- 1933: FM radio; first drive-in theater
- 1934: First tape recorder (magnetic tape)
- 1935: Canned beer; radar

- 1936: Colt revolver
- 1937: First jet engine; photocopier; typhus vaccine
- 1938: LSD synthesized
- 1940: Jeep; color TV (just in time for World War II)
- 1941: First computer using software; first program-controlled robot
- 1943: Term "cybernetics" coined by Julian Bigelow
- 1944: Discovery of DNA as the genetic material within each cell
- 1945: Atomic bomb; first public influence vaccine (flu vaccines were tested during WWII on the military); transponders (forerunners of RFID tech) used by military
- 1946: Microwave
- 1947: Roswell crash; holography invented; transistor
- 1950: 1:1 pairing of nitrogenous bases (adenine/thymine, guanine/cytosine) is proven; videotape recording; Turing Test is established as means to measure machine intelligence; Asimov publishes "Three Laws of Robotics"
- 1951: Rosalind Franklin and Maurice Wilkins use X-ray diffraction to show that DNA is helical; first working "artificial intelligence" program
- 1952: Patent for barcodes; hydrogen bomb (thanks to Edward Teller); polio vaccine
- 1953: James Watson and Francis Crick determine molecular structure of DNA
- 1953: Transistor radio
- 1954: Oral contraceptives; solar cell; anthrax vaccine

- 1955: Fiber optics
- 1956: First computer hard disk; term "artificial intelligence" first used
- 1958: Modem; laser; integrated circuit; Teddington Conference on the Mechanization of Thought Processes
- 1959: Microchip; Artificial Intelligence (AI) lab founded at MIT
- 1960: Book, *Man-Computer Symbiosis*, published by Licklider
- 1961: Valium; AI program solves calculus at college freshman level
- 1962: First computer game; oral polio virus; first industrial robot company Unimation founded
- 1963: Videodisk; measles vaccine
- 1964: BASIC (computer language)
- 1965: NutraSweet; compact disc; ELIZA (conversational computer program) successfully tested
- 1967: Handheld calculator
- 1969: Arpanet (forerunner to Internet); ATM; artificial heart; barcode scanner; Shakey the Robot successfully combines animal locomotion, perception, and problem solving; first International Joint Conference on Artificial Intelligence meets at Stanford
- 1970: Daisy-wheel printer; floppy disk
- 1971: Microprocessor
- 1972: Word processor
- 1973: Gene splicing; Ethernet; first recombinant organism

- 1974: Chicken pox vaccine
- 1975: Laser printer; first scientific publication by a *computer* (Meta-Dendral learning program)
- 1976: Ink-jet printer
- 1977: Magnetic Resonance Imaging (MRI); two new companies form—Amtech and Identronix (both arising from Los Alamos Labs research) to explore "civilian uses" for RFID (Los Alamos had developed the system to keep track of nuclear materials [for all the good it did])
- 1978: Vaccine for meningitis; MOLGEN program (written at Stanford) demonstrates object-oriented representation of knowledge that could be used in "gene cloning"
- 1979: Cell phones; Cray supercomputer; Sony Walkman; RFID transponders tried in dairy cattle; INTERNIST-I (medical diagnostic computer program)
- 1980: Hepatitis B vaccine; Supreme Court rules that genetically altered organisms can be patented (Diamond v. Chakrabarty); first conference of the American Association for Artificial Intelligence
- 1981: DOS; IBM PC; scanning tunneling microscope; RFID tried by railroads to keep track of "rolling stock"
- 1982: Human growth hormone genetically engineered (Humulin); RFID use begins in tollbooths and to identify fleet vehicles
- 1984: Apple Macintosh computer; RFID tags move from write once to programmable up to 500k times.
- 1986: Synthetic skin; superconductor; field tests of GMO tobacco

- 1987: Field testing on GMO tomatoes; Marvin Minsky publishes *Society of the Mind*, in which he claims the mind is a network of cooperating "agents"
- 1988: Digital cell phones; RU-486 (abortion pill); Prozac; Doppler radar
- 1989: HD Television
- 1990: World Wide Web; data mining begins
- 1992: FDA says GMO foods are not "inherently dangerous"
- 1993: Pentium Processor; Polly, the artificial intelligence bot that performs at animal speeds and uses "vision" to navigate
- 1994: HIV protease inhibitor; France approves GMO tobacco; two robotic cars navigate Paris traffic with passengers on board
- 1995: Java computer language; DVDs; ALVINN (semi-autonomous robot) steers a car across US (throttle/brakes controlled by human)
- 1997: "Deep Blue" chess computer defeats world champion Garry Kasparov
- 1996: WebTV
- 1999: Sony's AIBO dog becomes sensation as first autonomous robotic "pet"
- 1998: Viagra; Lyme disease vaccine
- 2001: Artificial liver; iPOD

The eighteenth century began with horse-drawn carriages, kerosene and/or whale oil lighting, and a narrow understanding of biology. In fact, little had changed from previous centuries. However, by 1899, gas or electric lighting had turned night to day, coal heating had been replaced with natural gas, automobiles shared the road with horses, man took flight in air balloons, and genetics had emerged as a specialty within the burgeoning discipline of biology. The rate at which science and industry announced new discoveries and invention in the twentieth century outpaced the previous century like a hare racing past a befuddled tortoise.

Building upon the racially biased, biosocial "psychometrics" of Sir Francis Galton, high-society pseudoscientists morphed genetics into eugenics (Greek for "true birth") in a self-serving bid to "improve the human condition." By 1900, Darwin's ideas permeated social science with racist rancor. David Starr Jordan, first president of Stanford University, authored a book in 1902 that distilled and codified the rising field of eugenics. I refer to *Blood of a Nation: The Study of the Decay of a Nation through the Survival of the Unfit.* (See this link to read the pdf version: http://www.nazi.org.uk/ eugenics%20pdfs/BloodOfTheNation.pdf). In this treatise, Jordan advocates a program of "artificial selection," in which inferior forms are destroyed and superior forms encouraged. But he went even further:

> To select for posterity those individuals which best meet our needs or *please our fancy*, and to destroy those with unfavorable qualities, is the function of artificial selection. Add to this the occasional crossing of unlike forms to promote new and desirable variations, and we have the whole secret of selective breeding. This process Youatt calls the *"magician's*

wand" by which man may summon up and bring into existence any form of animal or plant useful to him or *pleasing to his fancy*. (pp. 13–14 emphasis added)

Ultimately, Starr argued, it is war that weakens any society, for nations inevitably send their best to the battlefield, forever removing superior blood of those who die.

By 1910, scientists like David Starr Jordan and Charles Benedict Davenport joined ranks with America's northeastern elite (Harriman, Rockefeller, and Carnegie) to form the Eugenics Record Office (ERO) of Cold Spring Harbor, New York. The stated goal of the ERO was to study human patterns of heredity in order to eliminate the substandard or socially inadequate. To achieve this goal, the ERO sought out politicians across the country, advocating and even lobbying for enactment of sterilization laws that would prevent undesirables from reproducing. The ERO also promoted selective breeding programs that would improve the human stock, yielding greater percentages of the strong and bright.

The ERO's "advisory committee" included experts in statistics, criminology, physiology, biology, thremmatology (scientific "breeding" of selected plants and animals), history, law, religion and morals, anthropology, psychiatry, sociology, and even an oddly named "Woman's Viewpoint" offered by one Caroline B. Alexander.

The 1914 ERO Report ("Report of the Committee to Study and to Report on the Best Practical Means of Cutting Off the Defective Germ-Plasm in the American Population," compiled by H. H. Laughlin, a former high school teacher with a profound interest in Mendelian inheritance, available online, last accessed June 13,

2013: http://dnapatents.georgetown.edu/resources/Bulletin10A. pdf) divided the undesirables into ten classes:

1. The Feeble-minded Class
2. The Pauper Class
3. The Inebriate Class
4. The Criminalistic Class
5. The Epileptic Class
6. The Insane Class
7. The Asthenic Class
8. The Diathetic Class
   a. Species Difference
   b. Racial Difference
   c. Family and Individual Differences
9. The Deformed Class
10. The Cacæsthetic Class

Eugenics: the science that deals with the improvement of the races and breeds especially the human race through the control of hereditary factors

According to Laughlin, the purpose of the Eugenics Record Office is to determine a means to uproot and remove defective germ plasm from American inheritance. Laughlin refers to this as the "negative side of the problem," leaving the "positive side" (that of breeding better Americans through artificial selection) to others. Referring to those in the above list, Laughlin writes: "If they mate with a higher level, they contaminate it; if they mate with the still lower levels, they bolster them up a little only to aid them to continue their own unworthy kind. They constitute a breeding stock of social unfitness" (ibid., 15).

These early decades of the twentieth century formed a hideous breeding ground for the pseudoscience of "eugenics." Moral decay painted cities with a broad and hideous brush. The rise in immigration, particularly to America's eastern seaboard, led to perceived racial divides while segregating people groups within the confines of slum housing. Against this impoverished background glittered a rising aristocracy with new money and old family ties. The industrial age brought railroads, oil, and electricity. Henry Ford's assembly lines replaced craftsmen, creating affordable goods for the working man. Electricity bedazzled the modern streets of major cities, and soon dirigibles and prop-engine planes dotted the evening skylines. The twentieth century promised more products with less effort, and "modern living through chemistry" fireproofed homes; replaced dangerous glass with easy, unbreakable plastics; and even improved cosmetics.

Not every aspect of this brave, new world was rosy, however. War built new corporations, but it also filled furrowed fields with the blood of nations. As soldiers traveled across borders, H1N1 (dubbed the "Spanish Flu" because it first rose to international attention via an epidemic in Spain) decimated families across the globe. Vaccines and antibiotics changed medicine, but they also inoculated millions with SV40, a virus native to Rhesus monkeys whose livers had been extracted (posthumously) and used to grow Merck's polio vaccine in the 1950s and 60s. Now, SV40 is known as a carcinogen. This DNA virus plays havoc with a cell's natural lifespan, switching off a gene known as *TP53*, which encodes for the protein p53. The protein p53 plays a major role in tumor suppression by acting like a "stop signal" when a cell has reached the end of its life. This "stopped" cell would no longer divide—indeed it would die. So, SV40 appears to mutate *TP53*, changing the resulting protein's shape, which renders it incapable of acting as a "stop signal." The polio vaccine is not the only medicine with death in the mix. A statement by Dr. Maurice Hilleman, former chief of Merck's vaccine division, not only reveals SV40 as a component of the polio vaccine, but also appears to indicate the presumed unintentional inclusion of HIV in the hepatitis B vaccine. (See: http://www.youtube.com/watch?v=edikv0zbAlU.)

World War I never truly ended politically. Germany's defeat and humiliation served as a fecund surrogate for the birth of the world's next Napoleon: Adolf Hitler. Though most have read or heard of Hitler's drive to create an Aryan super-race of men (—*translated roughly as*), few realize that Hitler's inspiration derived from American and British eugenics programs, including twin studies performed and advocated by the Eugenics Records Office. According to Edwin Black, in an article published by the *San Francisco Chronicle* in 2003, "The concept of a white, blond-haired, blue-eyed master Nordic race didn't originate with Hitler. The idea was created in the United States, and cultivated in California, decades before Hitler came to power. California eugenicists played an important, although little-known, role in the American eugenics movement's campaign for ethnic cleansing" (Edwin Black, "Eugenics and the Nazis—the California Connection," *San Francisco Chronicle*, November 9, 2003; viewable online, last accessed June 13, 2013: http://www.sfgate.com/opinion/article/Eugenics-and-the-Nazis-the-California-2549771.php).

Believing his nation had become weak because of inferior bloodlines and degenerate breeding, Hitler employed both negative and positive approaches (as mentioned earlier by Laughlin in his report for the ERO) to improve Germany's bloodlines. Physicians were required to report any and all degenerative illnesses to the government. Hereditary "health" courts sprang up all over the country, leading to the forced sterilization of over four hundred thousand people (Robert Proctor, "Racial Hygiene: Medicine Under the Nazis," Cambridge, Massachusetts: Harvard University Press, 1988, 108). Those deemed so inferior that they presented a burden on society faced the ultimate penalty at the *Hartheim Euthanasia Centre*, where patients breathed their last in rooms filled with carbon monoxide. The *Lebensborn* program sought to increase pure Aryan numbers through extramarital affairs between SS officers (whose bloodlines were verified) and equally pure German females. The offspring of these affairs were raised by the state and placed with genetically acceptable families. Buxom, blond German women mated with muscular, intelligent blond men provided true Aryan genes to select for superior Aryan children.

Most historians would say that the eugenics era ended in 1979, when California's lawmakers struck down that state's longstanding, compulsory sterilization law. However, I would argue that the program continues in the drive to catalog the human genome and improve it. The Human Genome Project (HGP) provides the baseline that genetics researchers can now "tweak." The twentieth century brought an explosion of genetics discoveries. We've come a long way from the early days when David Starr Jordan sought to improve the "germ plasm."

According to the May 2011 report prepared by the Battelle Technology Partnership Practice, the project not only created three hundred and ten thousand private-sector jobs, but it also turned the initial $3.8 billion investment into a whopping $796 billion! Aren't you grateful for the long hours that the HGP scientists and techs spent slaving over lab benches? I know I am! Thanks to these countless hours, scientists now have reference information for 3 billion base pairs, dropkicking science into a brand new age!

Here's what the report mentioned above has to say about this brave new age of genetics:

> Scientists are using the reference genome, the knowledge of genome structure, and the data resulting from the HGP as the foundation for fundamental advancements in medicine and science with the goals of preventing, diagnosing, and treating human disease. Also, while foundational to the understanding of human biological systems, the knowledge and advancements embodied in the human genome sequencing, and the sequencing of model organisms, are useful beyond human biomedical sciences.
>
> The resulting "genomic revolution" is influencing renewable energy development, industrial biotechnology, agricultural biosciences, veterinary sciences, environmental science, forensic science and homeland security, and advanced studies in zoology, ecology, anthropology and other disciplines. (Battelle Technology Partnership Practice, "Economic Impact of the Human Genome Project," May 2011; viewable online, last accessed June 14, 2013: http://www.battelle.org/publications/humangenomeproject.pdf)

The baby boom generation, born to returning WWII veterans, has seen the world shift on its axis. In 1945, the war in the Pacific ended with the destruction of Hiroshima and Nagasaki, Japan, ushering in the nuclear age and a frantic study of genetic mutations by radiation. The war also ushered in unparalleled economic growth in the United States. Returning veterans found well-paying jobs, married, and built new homes. America's gross national product skyrocketed as middle-class numbers swelled. White color, service-industry jobs increased as our country shifted from producing goods to consumerism. The "service industry" replaced traditional factories. Corporations gobbled companies, and conglomerates fed on corporations. Thanks to Eisenhower's interstate system, goods moved easily and cheaply while American tourists fell in love with station wagons and "mobile homes."

In 1953, when Watson and Crick unraveled the structure of the DNA molecule, most children played with dolls or popguns. Television had only just emerged as a new form of entertainment, but the new medium had already become an integral part of most "boomer" families. Frozen dinners consumed while watching *I Love Lucy* or *Arthur Godfrey* had begun to replace home-cooked meals around the dining room table. New home appliances helped women finish housework and food preparation quickly, giving them more time to work outside the home. The postwar baby boom did far more than just create a new, massive US generation; it also created an insidious new lifestyle that would eventually tear families apart.

In stark contrast to their grandparents' childhoods, today's three-year-olds are handling rudimentary computers and learning that guns are evil. Free speech is a thing of the past, but free love is encouraged and even taught to kindergarten children as their right as humans. Knowledge has most definitely increased—so much so that the sum total of human knowledge is now said to double every ten years. Yet, with all these "improvements," mankind has never been farther from God.

In fact, "mankind" itself is about to become an anachronistic term. Homo sapiens, the wonderful creation of a loving Father, is about to become extinct—or so George Church and Hugo de Garis would have us believe. The DNA molecule that Watson and Crick discovered during the early years of the boomer generation isn't good enough for self-directed evolution. Scientists no longer seek merely to understand God's marvelous molecule; they seek to replace it with a synthetic construct of man's design. A concept called "transhumanism" lies at the core of this drive to alter mankind's genome, and it is this hubristic notion—this deliberate jab into the eye of God—that may influence not only the mechanism within the coming Mark of the New World Order, but it may also increase its popularity with today's Generation Y.

chapter six

A Generation Ready for the Mark

Before we discuss the science behind transhumanism (a "scientific" goal that merely repackages eugenics' drive to "create a better human"), let's briefly look at Gen Y. Simply put, Gen Y children are "millennial babies," or the grandchildren of baby boomers. Boomer children are often referred to as "Generation X," so the "Y" babies are those born to the "X" children. Perhaps not coincidentally, X and Y chromosomes determine infant gender, a concept that has become deliberately blurred during the years since World War II. Respectively, XY and XX no longer define male or female. Transgendered individuals, or even those who simply choose to identify with a particular "sex" option, have convinced many schools, workplaces, and even politicians to protect their "choices." A child born as an XY male may choose to "identify" as a girl—even going so far as to join girls' basketball teams and shower in the girls' facility. God's design for distinct male and female roles has been corrupted into "fifty shades of gray."

Generation Y is sometimes called "the echo boom," as it is the largest generation since the postwar boomers. In fact, some statistics list Gen Y as numbering 80 million or more. Beyond this group is Generation Z—or those born after 1995. Today's media culture is already targeting this generation as rising consumers, customizing advertising to reach gender-neutral, tech-hungry, post-human individuals. Perhaps even more than their parents, Gen Y youth look to genetics to take the next step in human evolution. What is that step? Consider popular culture and the memes Gen Y and Gen Z young people are exposed to every day. Graphic novels, films, television, and video games constantly and consistently teach that mankind is about to "level up." Young men and women (and those in between) can become werewolves, vampires, super soldiers, and spider men. The old human paradigm of two genders and five senses can be replaced with an infinite number of gender choices (or none at all!) and extra senses that allow the transhuman to hear colors, smell sounds, or tune into the very fabric of space and time. The biblical promise that we will "rise up as eagles" is twisted into promising wings that permit the new human to take physical flight. Generations Y and Z could live forever. At least, that is the lie they are being told.

Cybernetics and synthetic genomes will soon merge into the sexy Cylon of *Battlestar Galactica*. The post-human will be beautiful, perfect in form, and capable of superhuman feats. And he/she will live forever. "Ye shall be as gods" is the oldest lie, spoken to Adam and Eve by a garden snake long ago, but it is echoing once again in the promises to a generation of lost, unhappy youths. The Antichrist will play upon this inner need for perfection, this constant yearning to become a god. The youth of Hitler's day believed they would join their leader in the coming thousand years of Aryan rule over the Earth. Is it so far-fetched to say that coming evil one, foreshadowed by Adolf Hitler, will also promise his followers a plum part in his kingdom?

Remember this important fact: The Antichrist will claim to BE CHRIST. He will arrive as a peacemaker—as the answer to all of our worldly problems. It is also quite possible that he will claim to be an enhanced—or "trans"—human. It might be more accurate to call him a "trance" human, for the Antichrist will cast a spell upon today's youth, luring all those who do not call the true Christ their Savior to take his special Mark. This might be a chip inserted into the body used to track each person's location, but it may be *much* more. This Mark of the Beast might actually rewrite DNA.

Read/Write/Genomes

Many of you may remember the early days of computers and computing power. Commercially available computers arose about the same time as Watson and Crick's hallmark discovery. The US Census Bureau received the first UNIVAC computer on March 31, 1951, for a cost of nearly $1 million. In all, the government purchased forty-six UNIVAC computers, which is remarkable considering that each machine required its own small room! Back then, computing power for these monsters paled in comparison when measured against the tiny computer inside your cell phone.

In just fifty years, computer technology has advanced incredibly fast, solid-state hard drives have replaced old magnetic tape, and chips have replaced vacuum tubes. It's very likely that the computer in your hand today is flash-based, containing no hard drive at all. Cloud computing will soon mitigate the need for "local storage," although (for now) you can still buy laptops with more than a terabyte of storage! One of the first computers one of these authors had was an old Commodore 128 that had 128 kilobytes of RAM. The laptop I'm using this morning has 6 gigabytes of RAM.

Hard-disk storage—in the old days—consisted of ROM, or "Read-Only Memory," which held chips or areas of storage that included operating software that must remain permanently available but rarely or never required updating. As computers advanced and "floppy disks" were added, most of these disks were "write once," which meant data could be written to them only one time. They could not be rewritten.

At one time, scientists considered our DNA as "Read-Only Memory." The code within each of our cells contains instructions for making the machinery that runs the complex operations that keep us alive. However, beginning in the twentieth century, early geneticists discovered a process called recombination that allows the insertion of new genetic material into a cell's genome. Recombination makes "gene therapy" possible, using a bacterial or viral vector, or "truck," to deliver the new genetic sequence to the target cell (perhaps in the lung or muscle). Scientists see nothing ethically wrong with "editing" our God-given genome; they merely see this as another form of self-directed evolution. Of course, one can only begin to edit what one reads first. Enter the Human Genome Project (HGP).

It might be said that the HGP actually began on December 9, 1984, when a group of scientists gathered in the Alta, Utah, ski resort area high up in the Wasatch Mountains. There, David Smith of the Department of Energy and Mortimer Mendelsohn of Lawrence Livermore Laboratory gathered with seventeen genetics scientists to determine whether or not the time might be right to commence serious measurement of the mutations within the human genome. Joining Smith and Mendelsohn were David Botstein, Ebert Branscomb, Charles Cantor, C. Thomas Caskey, George McDonald Church, John D. Delaharty, Charles Edington, Raymond Gesteland, Leonard Leman, Michael Gough, John Mulvihill, Richard Myers, James V. Neel, Maynard Olson, Edwin Southern, Sherman Weissman, and Raymond L. White.

Of this stellar list of respected names, one stands out: George McDonald Church. Church is a self-professed transhumanist and author of the recent best seller, *Regenesis: How Synthetic Biology Will Reinvent Nature and Ourselves*. He's a dedicated vegan, an entrepreneur (having established nine genomics-based businesses, including *Knome*, *Pathogenica*, and *Gen9Bio*—a synthetic biology company), and an avowed transhumanist who would love to see revived Neanderthal, assuming a human female volunteers to be the surrogate mother. (Oh, and by the way, Church was also instrumental in kick-starting the "race for the brain," discussed later in this section.)

Initially, the Alta group had discussed finding a repeatable and affordable means to quantify mutations within the Hiroshima offspring. Two years later, in 1986, Mendelsohn joined several others from Alta, as well as panel chair Arno Moltulsky of the Center for Inherited Disease, to present a report called "Office of Technology Assessment, Technologies for Detecting Heritable Mutations in Human Beings." This report had followed a meeting of leading molecular biologists in the spring of that year organized by James Watson (yes, that Watson), and held at Cold Spring Harbor Laboratory. Cold Spring, if you remember your eugenics history, was the site for the Eugenics Records Office (ERO). Are you beginning to get a nagging twitch at the back of your neck? Me, too.

By 1987, promotion of a planned genomic sequencing event kicked off with an article in the nation's most prominent newspaper, *The New York Times (NYT)*, titled "The Genome Project":

IT WOULD BE THE BIGGEST, costliest, most provocative biomedical research project in history, and the United States must embark on it immediately. That was how Walter Gilbert, Nobel Prize-winning biology professor at Harvard University, heard the genome project described at scientific meetings all through 1985 and 1986. The undertaking—which would reveal the precise biochemical makeup of the entire genetic material, or genome, of a human being—would, he heard, revolutionize medicine. It would answer the Japanese challenge in biotechnology. It would **grant insight into human biology previously held only by God.** (emphasis added)

After fawning over several scientists who attended the Alta meeting, the *NYT* article continues:

Today, a new consensus is emerging. Robert Cook-Deegan, an analyst at the Office of Technology Assessment (O.T.A.), a Congressional agency, reports that of late, "N.I.H. has been talking to D.O.E. There's more cooperation than friction." A bill has been introduced in Congress by Senator Pete Domenici of New Mexico to create a Government consortium to map and sequence the genome. Other bills are in the offing. Few doubt that the genome project, in some form, will eventually get under way.

Nobel laureate Walter Gilbert, though you may not know it, partnered with George Church in the study of mouse and yeast genetic elements. Gilbert also cofounded the company Biogen (now Biogen Idec) in 1978. Wally Gilbert may not be an avowed transhumanist, but he does spend a lot of time with Church, so one wonders if he might not be influenced by such proximity. Church is an evangelist when it comes to transhumanist dogma; perhaps it's discussed over a nice plate of organic sprouts.

With the *NYT* and other media moguls on board and a bill before Congress, it didn't take long for the idea of a gene race to take hold. In 1989, before leaving office, President Ronald Reagan approved the plans for a US-led Genome Project and placed it in his 1988 budget. In May 1990, the official proposal went to Capitol Hill under the title *Understanding Our Genetic Inheritance, The U.S. Human Genome Project: The First Five Years (FY 1991–1995).* Don't let the title fool you; the plan was for a fifteen-year study of the human genome to commence in 1991.

On September 11 of that same year, President George H. W. Bush gave a speech in a rare appearance before a joint session of Congress to discuss the economy and the rising conflict in the Persian Gulf. On that auspicious occasion, Bush said this:

> As you know, I've just returned from a very productive meeting with Soviet President [Mikhail] Gorbachev, and I am pleased that we are working together to build a new relationship. In Helsinki, our joint statement affirmed to the world our shared resolve to counter Iraq's threat to peace. Let me quote: "We are united in the belief that Iraq's aggression must not be tolerated. No peaceful international order is possible if larger states can devour their smaller neighbors."
>
> Clearly, no longer can a dictator count on East-West confrontation to stymie concerted United Nations action against aggression.
>
> A new partnership of nations has begun, and we stand today at a unique and extraordinary moment. The crisis in the Persian Gulf, as grave as it is, also offers a rare opportunity to move toward an historic period of cooperation. Out of these troubled times, our fifth objective—a **new world order**—can emerge: A new era—freer from the threat of terror, stronger in the pursuit of justice and more secure in the quest for peace. **An era in which the nations of the world, east and west, north and south, can prosper and live in harmony.**
>
> **A hundred generations have searched for this elusive path to peace,** while a thousand wars raged across the span

of human endeavor, and **today that new world is struggling to be born.** A world quite different from the one we've known. **A world where the rule of law supplants the rule of the jungle.** A world in which nations recognize the shared responsibility for freedom and justice. A world where the strong respect the rights of the weak.

This is the vision that I shared with President Gorbachev in Helsinki. He and the other leaders from Europe, the gulf and around the world understand that how we manage this crisis today could shape the future for generations to come. (emphasis added)

Is it a coincidence that the leader of the free world included such language in a speech that comes just as the world of science has initiated the most ambitious project of them all: to decode God's signature within our cells?

The slow and steady progress from eugenics to genetics to a New World Order is not coincidental, dear reader. It is part and parcel of an ages-old plan to unseat the true God and replace Him with a false idol, even man himself. Transcendence and transhumanism are the fruits of the Eugenics Record Office and the pioneering efforts of men like David Jordan and even our illustrious past president Theodore Roosevelt, who wrote this in a letter to prominent eugenicist Charles Davenport in 1913:

My dear Mr. Davenport:

I am greatly interested in the two memoirs you have sent me. They are very instructive, and, from the standpoint of our country, very ominous. You say that these people are not themselves responsible, that it is "society" that is responsible. I agree with you if you mean, as I suppose you do, that **society has no business to permit degenerates to reproduce their kind**. It is really extraordinary that our people refuse to apply to human beings such elementary knowledge as every successful farmer is obliged to apply to his own stock breeding. Any group of farmers who permitted their best stock not to breed, and let all the increase come from the worst stock, would be treated as fit inmates for an asylum. Yet we fail to understand that such conduct is rational compared to the conduct of a nation which permits unlimited breeding from the worst stocks, physically and morally, while it encourages or connives at the cold selfishness or the twisted sentimentality as a result of which the men and women ought to marry, and if married have large families, remain celebates [*sic*] or have no children or only one or two. **Some day we will realize that the prime duty—the inescapable duty—of the good citizen of the right type is to leave his or her blood behind him in the world; and that we have no business to permit the perpetuation of citizens of the wrong type.** ~~at all~~.

Faithfully yours,

Theodore Roosevelt

[And at bottom of left corner, the intended recipient:]

Charles B. Davenport, Esq.,

Cold Spring Harbor, L.I.

(emphasis added; Theodore Roosevelt, "Society has no business to permit degenerates to reproduce their own kind" letter, *Letters of Note*, posted with image of original letter on March 25, 2011, viewable here: http://www.lettersofnote. com/2011/03/society-has-no-business-to-permit.html)

Anyone reading today must be shocked to discover that our famous president actually believed it to be the duty of American citizens to leave behind the "right type" of blood! Adolf Hitler came to power based on an ideology steeped in opinions such as these, by taking the ideology of American eugenicists coupled with the financial backing of many within that movement, including George H. W. Bush's father, Prescott Bush!

According to a September 2004 article posted online in the UK paper, *The Guardian*, Bush's actions caused him to be indicted under the US "Trading with the Enemy Act":

George Bush's grandfather, the late US senator Prescott Bush, was a director and shareholder of companies that profited from their involvement with the financial backers of Nazi Germany.

The Guardian has obtained confirmation from newly discovered files in the US National Archives that a firm of which Prescott Bush was a director was involved with the financial architects of Nazism.

While there is no suggestion that Prescott Bush was sympathetic to the Nazi cause, the documents reveal that the firm he worked for, Brown Brothers Harriman (BBH), acted as a US base for the German industrialist, Fritz Thyssen, who helped finance Hitler in the 1930s before falling out with him at the end of the decade. The Guardian has seen evidence that shows Bush was the director of the New York-based Union Banking Corporation (UBC) that represented Thyssen's US interests and he continued to work for the bank after America entered the war. (Ben Aris and Duncan Campbell, "How Bush's Grandfather Helped Hitler's Rise to Power," *The Guardian*, September 25, 2004, http://www.theguardian.com/world/2004/sep/25/usa. secondworldwar)

Though the *Guardian* had timed the publication of this article to coincide with Bush's reelection campaign, we can thank the dark world of politics for its publication! Because of this, you and I can make the connections necessary to help us discern the times in which we live. Since the dawn of the twentieth century, a dark force has been rising within the confines of this sleeping giant of a country. Eugenics and eventually neo-eugenics and transhumanism have dictated scientific direction and set goals for politicians, all with one end in mind: To achieve the New World Order, to give it birth. During the past century, knowledge has increased exponentially, and the Internet has enabled us to "travel" at the speed of light. We live in the final days of planet Earth as we know it. Our world is bombarded with chemtrails and electromagnetic waves. Our foods and animals have been genetically altered, the air poisoned with radiation and heavy metals, and my genes and yours are held together only by the grace of God.

The Human Genome Project, intended to finish in fifteen years, instead announced completion only *thirteen* years into the work.

If you read summaries and reviews of that work now, you'll find many arguing that the announcement came prematurely—that the entire sequence had not fully been read. Thirteen is a number repeated again and again in occult circles. It is considered the number of rebellion, an unlucky number. Osiris was torn into fourteen pieces: thirteen plus his reproductive "member" (reconstructed by Isis and used to impregnate herself and then give birth to Horus). Do a search in the daily news for the number thirteen, and you will be amazed at how many times it occurs. Could it be that the HGP leaders arbitrarily stopped at thirteen and "rested," announcing their victory to the world, or is it possible that the number was craftily chosen as a nod to spiritual influences and masters?

The announcement ushered humanity into the post-genomic era and engendered the question: Who owns our genes? According to the final Human Genome Publication (No. 12), published in February 2002, the "First International Conversation on Enviro-Genetics Disputes and Issues," sponsored by the Einstein Institute for Health and Science, met in Kona, Hawaii. The meeting took place in July 2001 and was attended by more than eighty judges and forty scientists—all there to discuss the legal and ethical ramifications of the Human Genome Project's results. These legal and scientific experts joined together from all over the world to dissect the New World Order of the post-genome era. Shortly thereafter, attendee Justice Artemio V. Panganiban of the Philippines presented a paper with the assigned topic, "Paradigm Shifts in Law and Legal Philosophy," in which he recounted some of the opinions expressed during the Kona conference. Should you desire to read the entire paper, it is available online here: https://www.google.com/url?sa=t&rct=j&q=&esr

c=s&source=web&cd=2&ved=0CDQQFjAB&url=http%3A%2
F%2Fwww.cdi.anu.edu.au%2FCDIwebsite_1998-2004%2Fph
ilippines%2Fphilippines_downloads%2FPhilJudgoba.
rtf&ei=Z9X7UdzmHM3eyAGn-YDIDQ&usg=AFQjCNHdH
uOC5jRN8G4hrzkgDrU-ZnO77A&sig2=nTF7Er4XNHji2TI-
1EMBig; page last accessed August 2, 2013).

Justice Panganiban's comments succinctly summarize those of
many globalists:

> I believe that the major transformational shifts in the world
> have been brought about mainly by the informational and
> technological revolution unfolding even now as I speak. I
> refer to computerization, minuterization [*sic*], digitization,
> satellite communications, fiber optics and the Internet—all
> of which, taken together, tend to integrate knowledge on a
> worldwide scale. This international integration of knowl-
> edge, technologies and systems is referred to as globalization.

Amazingly, Panganiban has here distilled this section so far in
that knowledge, travel, and commerce have increased exponentially
since the turn of the twentieth century. He goes on:

> Professor Anne-Marie Slaughter of Harvard Law School,
> says that modern judges should "see one another not only as
> servants or even representatives of a particular government
> or party, but as fellow professionals in a profession that tran-
> scends national borders"… Justice Claire L'Heureux-Dube
> of the Supreme Court of Canada—in a speech before the

"First International Conversation on Enviro-Genetics Disputes and Issues" sponsored by the Einstein Institute for Science, Health and the Courts (EINSHAC) and held in Kona, Hawaii on July 1, 2001—opined that it is "no longer appropriate to speak of the impact or influence of certain courts on other countries but rather of the place of all courts in the global dialogue on human rights and other common legal questions."

The immediate impact of statements like these is that the legal representatives that you and I elect, or those appointed by such elected representatives, appear to owe their allegiance not to the electors but rather to "each other" and to the globe at large—to the "New World Order." So while you and I may still wish to believe that the act of writing to one's congressman precipitates action in Washington, the sad but obvious truth is that globalism rules our world, condemning our meager participation to the scrap heap of idealism and Judeo-Christian ethics.

The Human Genome Publication No. 12 referenced above has much to say about the global impact of the HGP. At the Kona meeting, National Institutes of Environmental Health and Sciences Director Kenneth Olden "observed that the Human Genome Project's output presents a major societal challenge to use the new information and technologies **to improve the quality of human existence**. The scope of this challenge, he said, is expanded further when the question is asked about who will benefit most from these advances and who will bear the greater share of the risk" (emphasis added [note how this language echoes the eugenics movement's goals to improve mankind]; US Department of Energy Office of Biological and Environmental Research, "Countering Bioterrorism: DOE-Funded DNA-Based Technologies Track Identity, Origin of Biological Agents," *Human Genome News*, Vol. 12, Nos. 1–2, February 2002, last accessed August 2, 2013: http://web.ornl.gov/sci/techresources/ Human_Genome/publicat/hgn/v12n1/HGN121_2.pdf).

The HGP publication continues with this informative paragraph:

> Discussions that followed the plenaries were far ranging, generally going beyond the suggested topics of genetically modified (GM) foods and agriculture, bioscience and criminal jurisprudence, biological property, genetic testing, and human subjects in biomedical research. During these sessions, judges related how their nations' courts have managed science and technology issues and the problems they have encountered. As groups attempted to anticipate issues likely to arise in the next two decades, researchers in their turn offered opinions on the current state of the science as well as some forecasts of advances on the near horizon. (ibid.)

I don't know about you, but reading language like this makes me feel a bit like a sick patient whose doctor is discussing my healthcare with my family while simultaneously ignoring me. As humans, each one of us carries unique DNA that sets us apart, making us special and one of a kind to God. However, corporations, judiciaries, politicians, and scientists huddle around our hospital beds like vampires in a blood bank, eagerly using us to further a selfish agenda.

And speaking of selfish, let me tell you a bit about an article I recently read called, "The Selfish Gene." Published in January 2013 at the DailyBeast.com and written by Michael Thomsen, the article bemoans the failures of the Human Genome Project—and Walter Gilbert in particular—to live up to the hype of their original "Promethean" promises. Citing the work of Hillary and Steven Rose, authors of *Genes, Cells, and Brains: The Promethean Promises of the New Biology*, Thomsen wonders whether the HGP provided consumers anything beyond a better "shampoo":

> For the Roses, the signal image of this movement is Walter Gilbert, one of the lead scientists on the Human Genome Project, standing on stage promising the possibility of fitting the code for human life onto a CD-ROM. This is the model for science in the Petri dish of post-industrial capitalism, reductionist fantasies delivered from a PR platform meant to turn public excitement into investor support. Scientists begin with a fixed conclusion, hyperbolize the benefits of reaching it, and then spend large amounts of private and public money to reach it only to discover their original promises were impossible. The Human Genome Project began not with a question, but an answer that had to be substantiated in reverse.

It's clear from this statement that the Roses and Thomsen believe the HGP was a letdown to corporatism as well as to you and me. Maybe he's missed the point. The PR campaign works in that it does sell the American or global public on an idea, particularly one that requires a serious infusion of what we naively call "our tax dollars." These "selfish" promises haven't failed; they have succeeded beyond measure! The post-genomic era has altered legal decisions, provided the foundation to define "human," given science a road map for gene hacking and synthetic biology, and propelled the transhumanist agenda to the forefront. Failed? Not on your tintype!

The New World Order, hailed as having arrived by President Bush in 1991, is here, and science has established its paradigm: a drive toward a new human…a *better* human. It's like the introduction to the hit 1970s TV program, *The Six Million Dollar Man*—"Gentlemen, we can rebuild him. We have the technology. We have the capability to build the world's first bionic man. Steve Austin will be that man. Better than he was before. Better, stronger, faster." Well, the post-genomic, proto-transhuman is just that and a bag of chips. And if you drag your feet and refuse to let your children or grandchildren participate in the Golden Age of the Transhuman Being, then you'll be branded a heretic at best, less sentient and perhaps even not worthy of life at worst.

The Golden Age of Transhumanism is upon us, whether we like it or not. Just look at the films, television programs, and video games set before our children for constant consumption and indoctrination. According to *H+ Magazine*, the top ten "transhumanist" films are, counting down from ten to one (my comments in brackets):

10. Avatar 3D (2009)
[If you've not seen this one, rent the DVD. It is the blueprint for 2025, when transhumanists such as Ray Kurzweil

and Natasha Vita-Moore see us transitioning to "avatars" in preparation for full uploading to computers. The plot is simple and "green": Mankind has conquered space, but corporations are still greedy and continue to strip resources—in this case, from another planet. When his brother is killed, paraplegic Marine Jake Scully takes his place on a mission to Pandora, a world filled with "backward" but natively beautiful creatures. Parker Selfridge is the greedy head of the corporation raiding Pandora, and his goal is to wipe out the "Na'vi" (the natives). The plot exploits the young by indoctrinating them with the "green agenda" while planting desires to be "free" to live an eternal, perfect life in a beautiful garden inside a virtual world.]…

9. Gattaca (1997)

[A film about hacking the human genome to build better people. The title is taken from the building blocks of DNA: A for adenine, G for guanine, C for cytosine, and T for thymine.]…

8. The Terminator (1984)

[James Cameron directed this as well as *Avatar*. Arnold Schwarzenegger is a sentient machine from the future. Oddly enough, this film presents the possible consequences of what Hugo de Garis calls the coming "Artilect War," in which sentient machines rise up against humanity (both transhuman and human) in a battle for control of the planet.]…

7. The Matrix (1999)

Probably only a few thousand people in the world have ever read, or even heard of, Nick Bostrom's confounding Simulation Argument. But tens of millions of people have watched this movie and thus have learned something about the concept. Which pill would you take? [This statement is from the *H+* article and contains the impressions of the writer. Bostrom's Simulation Argument posits that we are living within a living computer. Here is the abstract from this paper: "This paper argues that *at least one* of the following propositions is true: (1) the human species is very likely to go extinct before reaching a 'posthuman' stage; (2) any posthuman civilization is extremely unlikely to run a significant number of simulations of their evolutionary history (or variations thereof); (3) we are almost certainly living in a computer simulation. It follows that the belief that there is a significant chance that we will one day become posthumans who run ancestor-simulations is false, unless we are currently living in a simulation. A number of other consequences of this result are also discussed" (Nick Bostrom, "Are You Living in a Computer Simulation?" *Philosophical Quarterly*, 2003, Vol. 53, No. 211, 243–255; viewable online, last accessed August 2, 2013: http://www.simulation-argument.com/simulation.html).]

6. WALL-E (2008)

[This film teaches children and adults alike that robots feel, love, and even "mate." Wall-E is a garbage robot whose solitary job is to clean up the mess on Earth left behind by the humans (who now live off-world). "Eve," a sleek female robot, is sent by the out-of-condition humans (one might say "devolved") to see if Earth can once again sustain life.]…

5. Dr. Jekyll & Mr. Hyde (1931)

Among the movies I've listed here, this is the one that probably the fewest of our readers have seen. That's a shame, because it is truly a great film and it also addresses important issues for transhumanists to consider. What makes up the human personality? How do we define "good" and "evil" and who gets to choose the accepted definition? [Again, the statement here is from the article's author at *H+*, so it's interesting that he chose this old film, based on a nineteenth-century novel. *The Strange Case of Dr. Jekyll and Mr. Hyde* was published in 1886 by Robert Louis Stevenson. Many films, a play, and even a musical have been based on this novella. The notion of good and evil is also a biblical one.]

4. Eternal Sunshine of the Spotless Mind (2004)

In my opinion as a cineaste, this is the best movie made, of any genre, thus far in the 21st century. And it's also an essential film for those who are interested in the questions of neuroethics. We may not be far away from having technologies that can enable the precise manipulation of memories, and, by extension, of personality. What a treat when a wonderfully entertaining movie also can engage the viewer in a challenging exploration of transhumanist ethics. [The writer's enthusiasm for this film is telling. The title is taken from a poem by Alexander Pope about a woman whose comfort after losing love is forgetfulness. In the film, Jim Carrey's character Joel Barrish and his lover, Clementine (Kate Winslet), both choose to have their memories erased by the Lacuna Corporation. The story line follows the disappearing memories as they are erased inside Barrish's head with the final memory, "Meet me in Montauk." Many of you are already thinking what I'm about to write: Montauk, New York, is the site of the notorious "Montauk Project," related to the Philadelphia Experiment.]

3. Brazil (1985)

Total movie magic…and although most of the technology depicted in this film is steampunk, not *H*+, it should provide a wake-up call for us to be aware that the struggle for political power and social control is ever present, and that if we fail to pay attention, our dreams of techno-transcendence may be snatched away from us just when they seem within our grasp. [What an interesting comment! "Dreams of techno-transcendence"? May the Lord grant that you and I may continue to stand in the way of such "transcendence"!]

2. Metropolis (1927)

…In making *Metropolis*, the great director Fritz Lang used all the resources of the then world-class UFA studio in Berlin—nearly bankrupting it in the process—and achieved effects that have never been surpassed. See it on the big screen, preferably in the most recent restoration, and be blown away. [This film is in German with subtitles, but the restored copy is indeed beautiful to watch. The story takes place in 2026. Metropolis is modeled upon Babylon, and features apocalyptic themes, including a false prophetess called the "Whore of Babylon," who incites the workers to their doom. Robotics and human/machine evolution themes abound, as does the theme of class distinction, the wealthy versus the common worker. It's no wonder that the Germans loved it.]

1. 2001: A Space Odyssey (1968)

You knew we would end up here, didn't you? Where else but with a movie that not only is central to the concerns of transhumanists, but also is among the top ten or twenty movies ever made, of any kind. First to impress is its amazingly realistic depictions of life in space, whether on the shuttle—that flight attendant who walks upside down!— the space station, or the flight to Jupiter. And no film has ever depicted with such poignancy the troubled relationship between sentient AI and its human masters. Then comes the astonishing climax—the famous Stargate sequence— which I regard as Clarke's and Kubrick's attempt to portray, in cinematic terms, the human experience of a technological Singularity. Remember, this is decades before Vinge wrote his seminal essay, though in 1965 I.J. Good had first described the possibility of an "intelligence explosion," which likely influenced the filmmakers. Someday, perhaps, another even better movie will be made about the dreams/nightmares of transhumanists, but for now, this one is the pinnacle. [The writer effervesces over this one, but he has reason to do so. The movie *2001: A Space Odyssey* uses psychedelics and German music—"Also Sprach Zarathustra"—to paint a picture of mankind's directed evolution via panspermia, taking us from the deliberate intervention during the days of Neanderthal to a team of scientists in a spaceship and beyond. Nietzsche's coming *Übermensch* or Superman is revealed at the end—savior of all mankind in the transformed "starchild."] (Mike Treder, "Top Ten Transhumanist Movies," *H + Magazine*, November 8, 2010, last accessed October 3, 2013: http://hplusmagazine.com/2010/11/08/top-ten-transhumanist-movies/)

Honestly, how many of the above films have you seen? If you're not naturally a science-fiction fan, then it's likely that you've missed one or perhaps even all of them. However, these films represent what transhumanists consider core theology in compact, easy to digest, two- to three-hour lessons.

Fiction, and particularly fiction in film, provides immediate access to layers of the brain that mere nonfiction such as this cannot reach. When processing fiction, humans naturally "shut off" their logical tendencies to question a particular viewpoint. Taught the same viewpoint in the context of a pleasant or intriguing story, the "lesson" seeds itself into our subconscious and slowly begins to grow there.

Television has brought us loads of such "seeded" themes over the years since its invention. Early "space" programs appear silly to us now, but what 1950s child didn't long to be either Superman or Flash Gordon? However, beginning with a program in 1966, everything changed. Gene Roddenberry spent his youth as a pilot in World War II, but upon returning to civilian life, he had an early glimpse into the brand-new medium called television, and his life and our futures took a dramatic turn. Gifted with the ability to write compelling fiction, Roddenberry moved to Hollywood and began churning out scripts for early favorites like *Have Gun Will Travel, Highway Patrol, The Kaiser Aluminum Hour, The Naked City,* and *Dr. Kildare.* In 1964, Roddenberry decided to write a speculative script for a new science-fiction drama that he described as "Wagon Train in Space." The program was called *Star Trek,* and Roddenberry struck a deal with NBC to air the show beginning with the fall lineup in 1966. Despite climbing ratings, NBC canceled the show after eighty episodes (about three and a half seasons). Fan loyalty proved everlasting, however, and soon *Star Trek* conventions revived interest in the Enterprise crew, so much so that Paramount Studios signed Robert Wise to direct the original cast in *Star Trek: The Motion Picture,* which premiered in December 1979—and the rest, so they say, is history.

But is it? There is a secret history to the Roddenberry story that you may not know. Author Peter Levenda included the legendary writer as one of the inside crowd around a nebulous but highly influential group called "The Nine." This is what symbologist and science-fiction researcher Christopher Loring Knowles, author of *Our Gods Wear Spandex: The Secret History of Comic Book Heroes,* had to say about Roddenberry on his weblog, "The Secret Sun," in 2008:

In early 1975, a broke and depressed Roddenberry was approached by a British former race car driver named Sir John Whitmore, who was associated with a strange organization called "Lab-9." Though unknown to the public, Lab-9 were ostensibly a sort of an independent version of the *X-Files*, dedicated to the research of paranormal phenomena. However, Lab-9 had another, more complex agenda—they later claimed to be in contact with a group of extraterrestrials called the "Council of Nine" or simply "The Nine", who had been communicating through "channelers" or psychic mediums.

The Nine claimed to be the creators of mankind, and had informed the channelers that they would be returning to Earth soon. Lab-9 had wanted to hire Roddenberry to write a screenplay based on the Council of Nine's imminent return. To help Roddenberry in his research, Lab-9 flew him out to their headquarters, located on a large estate in Ossining, NY. There, Roddenberry met and interviewed several psychics, and prepared the groundwork for his script.

Roddenberry wrote a script called *The Nine*, in which he fictionalized his experiences at Lab-9 and the message for humanity that the Council of Nine wished to convey. But Roddenberry's story focused more on his fictionalized alter-ego and his marital and financial worries than on The Nine themselves, and Lab-9 requested a rewrite. He handed the task of revising the script to an assistant, Jon Povill. In his revision, Povill posited that the hit sci-fi TV show that

Roddenberry's alter ego had produced in the 60s was not actually his work, but had been channeled through him by the Council of Nine. UFO cultists in the 70s and 80s would make similar claims about *Star Trek* itself....

It was later revealed in the 1977 book *Briefing for the Landing on Planet Earth* that The Nine claimed to be the figures whom the ancient Egyptians had based their Ennead, or pantheon of major gods, on. Another book of channeled messages from The Nine was published in 1992 and was titled, *The Only Planet of Choice: Essential Briefings from Deep Space.* However, little has been heard from The Nine since that book's publication. But it is worth noting that a year after it was published, a new *Star Trek* TV series appeared called *Deep Space Nine*. (Christopher Loring Knowles, "The Council of Nine and the *Star Trek* Pantheon," June 6, 2008, last accessed October 3, 2013: http://secretsun.blogspot.com/2008/06/council-of-nine-and-star-trek-pantheon.html)

Knowles may be onto something. In fact, once you start down the rabbit trail leading from transhumanism back to Hitler and to the eugenicists, you run through some rough terrain that includes Manson, The Nine, a group of "wandering bishops" with no real church affiliation, assassins, and—of all things—H. P. Lovecraft, whose wife (some say) had an illicit affair over some months with none other than Aleister Crowley!

Forgetting all that for a moment (if indeed, one can), let us return to The Nine. Levenda not only addresses this odd bit of exopolitical history, but he dedicates an entire section of Book One to it. Here is the crux of that tale, broken down for us by "Gordon" on RuneSoup.com in an article called "The Séance that Changed America":

The man at the centre of this séance [the one that changed America] was Andrija Puharich, US Army Captain and author of a government paper on the weaponisation of ESP. And this is the guy that is moving in the same murky circles as [wandering] bishops Jack Martin and Fred Crisman.

The farmhouse in question was owned by his bizarro Round Table Foundation [RTF]which appears to have received funding from the CIA.

Puharich first gathered together these nine people on a warm night in early June. But the most interesting results were actually achieved in New Year's Eve of the same year.

And it's a line-up that positively defines "could not make this up". The group included:

Arthur Young, who invented the Bell helicopter. However at the end of WWII he abandoned military aviation to concentrate full-time on the paranormal.

Arthur's wife, Ruth…previously of the Forbes dynasty. Her son, Michael, would get a job at Bell Aerospace through her and Arthur's influence. (Michael's wife got Lee Harvey Oswald his job at the book depository. She was learning Russian from Oswald's wife who was living with her in Irving, Texas. Oh, and her father worked for a CIA front called the Agency for International Development. Lee

Harvey Oswald left the coffee company in New Orleans, saying to his co-workers he was "going to work for NASA." After the assassination, two other coffee company employees get jobs at NASA. Just saying.)

Mary Bancroft; of the Bancroft dynasty who would much later sell the Dow Jones and *Wall Street Journal* to Murdoch. She also happened to be the mistress of the then-CIA chief. (The one JFK fired after the Bay Of Pigs after saying he was also going to break up the CIA…who conveniently went on to investigate JFK's death. Just saying.)

Marcella Du Pont of the Du Pont family.

Alice Bouverie who was born into the Astor dynasty. (Her father died on the Titanic and her first husband was a Czarist prince who would work for the OSS during WWII.) Here's what happened at the séance [quoting from Levenda]:

> These gods, who were nine in number as well, were part of one great, creator god known as Atum. The other gods consisted of Shu, Tefnut, Geb, Nut, Osiris, Isis, Seth, Nephthys, and sometimes Horus.
>
> Communication with these entities was handled by the medium, an Indian gentleman referred to as Dr. D.G. Vinod, who slipped into a trance state at 12:15 AM and began speaking as "The Nine" by 12:30. Afterwards Dr. Vinod would claim to have no memory of the conversation that preceded between the Ennead Nine and their human counterparts.
>
> During the course of the séance the mystical

Nine informed the human nine that they would be in charge of bringing about a mystical renaissance on Earth. From there The Nine ventured into quasi-scientific, philosophical constructs that eventually led to the acknowledgement that they, the Grand Ennead, were in fact extraterrestrial beings living in an immense spaceship hovering invisibly over the planet and that the assembled congregation had been selected to promote their agenda on Earth.

Not a bad collection of people to pull together if you wanted to promote a specific agenda over the second half of the twentieth century. Untold riches and connective power in one farmhouse. In fact, you have to wonder what percentage of American wealth was controlled by people related to the attendees.

Writing about the face on Mars and its relation to a descendant group sprung from this very séance, Chris Knowles points out:

> And the other conundrum here is if the Council of Nine's psychics saw this thing before it was photographed in 1976, did NASA go looking for it solely based on their advice? What does that say about the influence of a group that most people could be excused for dismissing as a bunch of gullible New Agers?
>
> The Nine would go on to surface in weird places

for decades including near Uri Geller (the AP is Puharich, who first brought Geller to the US), President Ford, Gene Roddenberry (Deep Space Nine anyone?), Al Gore as well as Soviets surrounding Gorbachev who were instrumental in the collapse of communism as mentioned in this old Fortean Times piece. It's not unreasonable to assume there were many more such places. (Gordon, The Séance that Changed America," *Rune Soup*, last accessed October 3, 2013: http://runesoup.com/2012/05/the-seance-that-changed-america/#ixzz2aqB89N5l)

Are you beginning to wonder what else you may have missed in history class? While it's possible that all this is a fabrication, much of it does ring true as Levenda, who is a meticulous researcher, documents his claims in the pages of his books.

How many young people, scientists, and politicians have been heavily influenced by the decisions made by those under the sway of The Nine? No one can say for certain, but if the accusations are true, then it's possible that nearly every war, every invention, and even every lesson taught in your child's school can be traced to this infamous hidden 33 (nine is three to the third power).

Just when did "The Nine" first emerge, and why? For answers, let's begin with a look at Andrija Puharich, parapsychologist, inventor, and founder of the Round Table Foundation. Puharich was born in 1918 in Chicago to poor, Yugoslav immigrants. His birth name was Karel, but his parents lovingly called him Andrija, a nickname that he later adopted (perhaps because it sounded more mysterious). A gifted student, Puharich received a scholarship to Northwestern, where he studied pre-med, going on to receive his medical degree from Northwestern's medical school in 1943. During this same period, the young Puharich served with the Army Medical Corps, training at Fort Detrick (coincidentally, Fort Detrick is also the location for the army's biological weapons program from 1943 to 1969, when the program name changed to "biological defense"). Puharich claimed that he lectured on parapsychology to military audiences and invented an implantable "tooth radio," which he sold to the CIA. In fact, Puharich's connections to the CIA and the mind also led to LSD, the drug that killed CIA scientist Frank Olson.

Puharich's connections to the military go much deeper. While working with psychic and spoon-bender Uri Geller at the Stanford Research Institute (SRI), Puharich was approached by the Department of Defense (DOD) regarding computer safety. In a speech to the Psychotronics Conference on Disease and Biological Warfare Control (New York, 1987), Puharich had this to say:

> An incredible but absolutely true scene took place when Uri Geller was working on one floor at Stanford Research Institute (SRI). They had Geller bending metal, teleporting things, demonstrating incidents of telepathy and clairvoyance—these things were happening all of the time. Well, unbeknownst to us at the time, there was another lab upstairs for ARPA—a computer network system. Somebody put two and two together and said: "Hey, there's a crazy kid downstairs who is bending metal and levitating

things." So they cross-correlated and discovered that when Uri did something the computers would go wacko: program printouts would pop out—sometimes partly erased—the power supply would go out on them and so on. "Somebody can affect the computer!" Panic ensued. **A squad of colonels came out from Washington to sniff around and watch Uri do his thing. They came to me and said, "You know, our whole defense system is on computers and magnetic tape cards. Can this guy wipe them out? Would you cooperate?" So we took Geller to Bell Labs and to the Livermore Radiation Lab and they put together an elaborate set-up for magnetic shielding. They learned that he could wipe out anything on computer tape. They said to me, "This guy could start World War 3!"** (emphasis added; Lawrence Gerald, "Surfing the E.L.F. Waves with Andrija Puharich," *Reality Hackers Magazine*, 1988, as quoted by *The Psychedelic Shakespeare Solution*: http://www.sirbacon. org/4membersonly/puharich.htm)

It's of particular interest that the SRI experiment listed above was reportedly under the control of former Apollo astronaut, Edgar Mitchell, who also sits on the board of the IONS (Institute of Noetic Sciences) as a cofounder. If one places Puharich at the center of a circle, the connections radiating out touch a vast and disturbing network of military, political, covert, scientific, and even entertainment fields! As mentioned earlier, we are being programmed by the media and arts with themes based upon governmental directives— directives that might be based upon orders from above (The Nine)!

Puharich considered the mind the most important "new frontier," and he even wrote a book on the topic called *The Sacred
Mushroom: Key to the Door of Eternity*, in which he described
how he had rejoined the military as a medical doctor in 1952
(his original stint had ended as a medical discharge based on a
recurring middle-ear condition). His second run with the Army
saw Puharich serving as a captain in San Antonio, Texas. One of
his first assignments came directly from the top, when Puharich
was "ordered" to give a lecture on ESP to the Aviation School
of Medicine. Shortly after this, Captain Puharich moved to the
Army Chemical Center in Maryland, where he served as the chief
of the outpatient clinic, but his next assignment took him back to
his former love: the study of the human mind. Here's how Puharich described that moment:

[I]n November of 1953 my colonel friend in the Pentagon
called me up one day and said that a way had been worked
out whereby the Army could sponsor my researches into
extrasensory perception. This was to be arranged through
a university which would act as a blind for the Army interest in this forbidden subject. After several months of nego-
tiation all this, too, came to naught. Therefore, I was more
than surprised to have the subject reopened by a responsible
officer, Colonel Nolton, on my own post, however circu-
itous and indirect the approach may have been. I told him
that in my opinion extrasensory perception was a reality,
and that it could be proven in people with exceptional talent. I pointed out that there was also evidence to the effect

that the talent was widely diffused throughout a normal population, and that it was probable that everyone has some of it sporadically.

"Well, if this is true," he persisted, "isn't it possible to find some drug that will bring out this latent ability so that normal people could turn this thing on and off at will?"

"It would be nice to have such a drug," I replied, "because then the research problems of parapsychology would be half solved. You see, the main problem in extrasensory perception research is that we know, even in a person of great talent, when this mysterious faculty will manifest itself. So we just sit around like a fisherman in a boat who puts his hand into the water every once in a while, hoping that a fish will swim into his grasp. There have been some reports of primitive peoples using such drugs extracted from plants, but I have never heard of one that worked when tested in the laboratory.

"Well, if you ever find a drug that works let me know, because this kind of thing would solve a lot of the problems connected with Intelligence." This was the parting word of the colonel as the conversation ended. (Andrija Puharich, *The Sacred Mushroom*, 8 of pdf, last accessed August 6, 2013; found at http://bearsite. yi.org/General/Philosophy/%2528ENG%2529%20 Puharich%252C%20Andrija%20%20The%20sacred%20 Mushroom%252C%20Key%20to%20the%20Door%20 of%20Eternity.pdf; also available here: http://wiki.zimbra. com/images/archive/2/2b/20110603152623!13936.pdf)

Puharich claims in his book, *The Sacred Mushroom*, that he discovered mushrooms could be used to induce trances while reviewing the text of "automatic writing" and channeled information given to him by one Alice Bouverie. Bouverie included the strange text with a letter explaining that Harry Stone had offered the cryptic information after being handed a gold pendant bearing the name of the Egyptian Queen Tiy. Here is how Bouverie described the moment to Puharich:

"Well, I had no sooner handed it to him than he trembled all over, got a crazy staring look in his eye, staggered around the room a bit, and then fell into a chair. I was petrified and really thought he was having an epileptic fit. Betty said that she had never seen Harry like this before. I rushed to get some water while Betty held him up. When I got back he was sitting rigidly upright in the chair and staring wildly into the distance. He didn't seem to see us at all but was watching something we couldn't see."

"Sounds as though he was in a trance, doesn't it?"

"Yes, that's what it turned out to be, but at that moment I had no idea what was going on. It probably wouldn't alarm a doctor like you, but I had never seen anyone go into a trance before."

"Well, what happened to make you believe that this was really a trance?"

"He just sat staring into nowhere for about five minutes. Then he jumped up and clutched my hand sort of desperately. I must say, it was awkward and embarrassing, especially because of the way he kept staring into my eyes. I have never seen such fanatically blue eyes in my life. He kept saying, 'Don't you remember me, don't you remember me?' over and over. And I kept saying over and over, 'Of course, Harry, I remember you.' But this made no impression on him. Then he began to speak quite clearly in English about his upbringing. I didn't realize that there was anything extraordinary about what he was saying until he asked for a paper and pencil and began to draw Egyptian hieroglyphs. I'm sure they were hieroglyphs, even though I don't know a word of Egyptian. This finally made me realize that he was in a somnambulistic state. Then he started to tell me about some drug that would stimulate one's psychic faculties. That is why I called you, because you're the only person I know who might be able to make sense out of what Harry said. I'd like to know what this is all about."

"Well, it sounds interesting. Why don't you send me a transcript of what he said, and I'll give you my opinion."

The statements given to Puharich by Bouverie "revealed" an ancient Egyptian method for inducing a state of anesthesia that also separated the mind from the body.

I turned to the material in paragraph four that Alice had sent me. The drawings made by Harry in trance certainly looked like mushrooms to my untrained eye. I must confess that at this time the world of mycology was virtually unknown to me. My only acquaintance with fungi was a half-hour lecture in medical school on the treatment of mushroom poisoning. I knew only that if a patient was brought in and suspected of mushroom poisoning, and that if certain symptoms were present, one used atropine as an antidote.

Since I had never had to treat a case of mushroom poisoning I can hardly say that I was sure of even this bit of information.

Remember that, according to Puharich's own account, his superiors wanted him to find a drug that could induce an ESP trance. The Egyptian mushroom drawn and described by Harry Stone led Puharich to the *Amanita muscaria,* used in Mexico as an intoxicating beverage and by Siberian shamans.

Excited by this discovery, Puharich arranged to meet with Harry Stone to test the sculptor's psychic abilities and to learn more about the mushroom. During this meeting, Captain Puharich blindfolded Stone before handing him the gold Egyptian pendant wrapped in cotton. As before, Stone stiffened and began to speak in the voice of an Egyptian known as Ptah Katu (below, "a. p." stands for Andrija Puharich; "h. s." for Harry Stone):

a. p. Ptah Katu, when you rubbed the plants, the white spots, did you mix them with something? (h. s. again indicated by a gesture something from a high tree.)

a. p. Tall trees? There was something that came from the tall trees? Palm trees? Coconut?

(h. s. nods in assent and now begins to rub the top of his head.)

a. p. Did you use it like that on the head? Did you use it only on your sick people? (h. s. shakes his head no.)

a. p. On the priests? (h. s. nods yes.)

a. p. Is that how Antinea was initiated? (h. s. nods yes.)

a. p. How was this done?

h. s. By opening the door. By stepping in.

And by leaving.

But it was only for them who know.

It would be dangerous to say everything one knows. Isn't it?

Puharich is convinced that Stone's trance is genuine.

"I don't think he's faking the trance; it seems to be the real thing. But what does it all mean—this Egyptian language, and the mystery about the mushroom? If we can get this language decoded we may get an answer. He doesn't say much of importance in English, and retreats from direct questions about the mushroom. I feel that Harry is under a powerful influence when he writes in hieroglyphic. He appears to be a machine responding to a master control. I wonder if he could be under intelligent control?"

This observation is very interesting in light of the RTF conversation that Puharich says took place in Maine. According to Jim

Keith, in his book, *Mass Control: Engineering Human Consciousness*, Puharich founded the RTF in 1948 in Camden, Maine.

Among Puharich's associates at the Round Table were War-ren S. McCulloch, one of the founders of cybernetics the-ory, who had worked at Bellevue Hospital in New York. McCulloch was an early advocate of electric brain implants and chaired conferences sponsored by the Josiah Macy, Jr. Foundation, a channel for CIA mind control funding. Another associate of Puharich was John Hays Hammond, said to have been Nicola Tesla's only student.

It gets stranger. According to Lynn Picknett and coauthor Clive Prince in their 2001 book, *The Stargate Conspiracy*, the RTF was actually funded by the Armed Forces Special Weapons Project (Lynn Picknett and Clive Prince, *Stargate Conspiracy: Revealing the Truth behind Extraterrestrial Contact, Military Intelligence and the Mysteries of Ancient Egypt*, Berkley Publishing Group, New York, 1999, viewable online, last accessed October 3, 2013: http://books.google.com/book s?id=2P6DBTX3MXUC&pg=PT167&dq=puharich+round+table &hl=en&sa=X&ei=XikBUuyDMKGyAGFo4AY&ved=0CEAQ6A EwAQ#v=onepage&q=puharich%20round%20table&f=false)!

The Round Table also received monies from private funding—which, according to Picknett and Prince, included former vice president of the United States and avowed Freemason Henry Wal-lace, who is also credited with incorporating the Great Seal (includ-ing the all-seeing eye) into the design for the dollar bill. In fact, well-known medium Eileen Garrett said in her autobiography that Wallace often visited the Round Table during 1949–1950 (this is referenced in the abovementioned book by Picknett and Prince).

Andrija Puharich appears to have enjoyed some very deep connections to military and covert psi operations, does he not? It is these connections that add weight and further mystery to the now infamous 1952 meeting with Hindu mystic "Dr. Vinod," who then went into a trance and channeled The Nine:

M calling: We are Nine Principles and Forces, personalities if you will, working in complete mutual implication. We are forces, and the nature of our work is to accentuate the positive, the evolutional, and the teleological aspects of existence. By teleology I do not mean the teleological aspects of existence. By teleology I do mean the teleology of human derivation in a multidimensional concept of existence. Teleology will be understood in terms of a different ontology. To be simple, we accentuate certain directions as will fulfill the destiny of creation.

We propose to work with you in some essential respects with the relation of contradiction and contrariety.

The whole group of concepts has to be revised. The problem of psychokinesis, clairvoyance, etc., at the present stage is all right, but profoundly misleading—permit us to say the truth. Soon we will come to basic universal categories of explicating the superconscious. (Mark Russell Bell Blog, *Case Study: Uri*, by Andrija Puharich, viewable online, last accessed October 3, 2013: http://metaphysica-larticles.blogspot.com/2012/06/uri-journal-of-mystery-of-uri-geller.html)

Mysterious enough for you? Later, The Nine claim that "they" had prepared for Puharich to meet Uri Geller. "It was us who found Uri in the garden when he was three. He is our helper sent to help man. We programmed him in the garden for many years to come" (ibid.).

The following paragraphs are also taken from the mentioned Mark Bell Blog essay about his meetings with Geller and how they related to The Nine. I quote them here to cement in your mind, dear reader, just how Puharich's experiments and his connections to military psi ops and even to implantable "radio" devices lead right into the final deception during the New World Order:

On December 2, 1971, Puharich and Uri were riding in a Jeep in the Sinai Desert when Uri whispered to him, "Our teacher said to us that he is going to appear as a red light that will look like a UFO." After spotting a red light in the sky, Puharich realized that while Uri indicated he saw the light the two soldiers accompanying them said they could not see it. Puharich wrote:

Did this red light have something to do with the voice? And what was the voice? A fragment of Uri's mind? A spirit? A god? Did the voice have any relationship to The Nine that had reached me so many years ago? The red light that followed our Jeep now seemed to be totally unlike what I had seen in the sky in Ossining, New York, in 1963 or Brazil in 1968.

On Dec. 5th the disembodied voice was quoted in a message that included the statement:

Andrija, you shall take care of Uri. Take good care of him. He is in a very delicate situation. He is the only one for the next fifty years to come. We are going to be very, very far away. Spectra, Spectra, Spectra: That is our spacecraft.

Andrija: "How far away is it?"

It is fifty-three thousand sixty-nine light-ages away.

On December 7, climbing up an embankment, they saw "a blue stroboscopic light pulsing at about three flashes per second." This event was one of many descriptions of sightings of unidentified flying objects. Puharich quoted Uri's response to the recent events.

"When I was out in that field, I realized for the first time in my life where my powers came from. Now I know for sure that they are not my powers. Oh, I know that I have a little bit of telepathy and psycho-kinesis—everybody has some. But making things vanish, and having things come back, and the red light in the sky in the Sinai, the blue light tonight, that is the power of some superior being. **Maybe it is what man always thought of as being God.**" (emphasis added; ibid.)

II Thessalonians 2: 1–12 says this:

Now we beseech you, brethren, by the coming of our Lord Jesus Christ, and by our gathering together unto him, That ye be not soon shaken in mind, or be troubled , neither by spirit, nor by word, nor by letter as from us, as that the day of Christ is at hand. Let no man deceive you by any means: for that day shall not come, except there come a falling away first, and that man of sin be revealed, the son of perdition; Who opposeth and exalteth himself above all that is called God, or that is worshipped; so that he as God sitteth in the temple of God, shewing himself that he is God. Remember ye not, that, when I was yet with you, I told you these things? And now ye know what withholdeth that he might be revealed in his time. For the mystery of iniquity doth already work: only he who now letteth will let, until he be taken out of the way. And then shall that Wicked be revealed, whom the Lord shall consume with the spirit of his mouth, and shall destroy with the brightness of his coming: Even him, whose coming is after the working of Satan with all power and signs and lying wonders, And with all deceivableness of unrighteousness in them that perish; because they received not the love of the truth, that they might be saved. And for this cause God shall send them strong delusion, that they should believe a lie: That they all might be damned who believed not the truth, but had pleasure in unrighteousness.

We'll leave the strange world of Andrija Puharich now and head back into the modern realm of strange science. Puharich developed an implantable radio device that could be hidden inside a tooth. Today, we no longer need to implant transceivers into teeth because we now have voice-to-skull transmission.

In the early 1970s, Dr. Joseph Sharp of the Walter Reed Army Institute of Research used a computer to control a radar transmitter such that for each time a human voice waveform changed from a peak to a valley, the radar transmitter sent out a single pulse, causing a single click to be heard by the test subject. Because these clicks were timed according to the human voice waveform, the test subject heard a voice, rather than a string of clicks. This has not been pursued, at least publicly, due to concerns about the effect of microwave signals aimed at a person's skull, but it does work. (Eleanor White, "What Is Voice to Skull?" *Targeted Individuals: Canada Wordpress*, June 24, 2010, last accessed October 3, 2012: http://targetedindividualscanada.word-press.com/2010/06/24/article-what-is-voice-to-skull/)

White is correct.

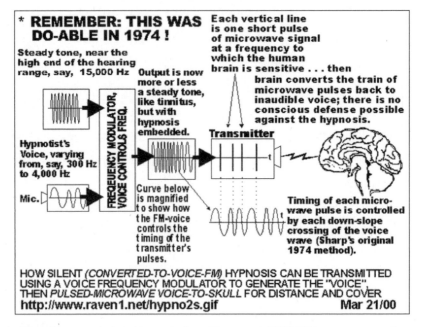

* **REMEMBER: THIS WAS DO-ABLE IN 1974 !**

Steady tone, near the high end of the hearing range, say, 15,000 Hz

Hypnotist's Voice, varying from, say, 300 Hz to 4,000 Hz

Mic.

FREQUENCY MODULATOR, VOICE CONTROLS FREQ.

Output is now more or less a steady tone, like tinnitus, but with hypnosis embedded.

Curve below is magnified to show how the FM-voice controls the timing of the transmitter's pulses.

Each vertical line is one short pulse of microwave signal at a frequency to which the human brain is sensitive . . . then brain converts the train of microwave pulses back to inaudible voice; there is no conscious defense possible against the hypnosis.

Transmitter

t

Timing of each micro-wave pulse is controlled by each down-slope crossing of the voice wave (Sharp's original 1974 method).

HOW SILENT *(CONVERTED-TO-VOICE-FM)* HYPNOSIS CAN BE TRANSMITTED USING A VOICE FREQUENCY MODULATOR TO GENERATE THE "VOICE", THEN *PULSED-MICROWAVE VOICE-TO-SKULL* FOR DISTANCE AND COVER

http://www.raven1.net/hypno2s.gif   Mar 21/00

(Above image source: http://www.wired.com/dangerroom/2008/05/army-removes-pa/)

In May 2008, Sharon Weinberger wrote this in her intriguing article, "Army Yanks Voice-to-Skull Devices":

The Army's very strange webpage on "Voice-to-Skull" weapons has been removed. It was strange it was there, and it's even stranger it's gone. If you Google it, you'll see the entry for "Voice-to-Skull device," but, if you click on the website, the link is dead.

The entry, still available on the Federation of American Scientists' website, reads:

Nonlethal weapon which includes (1) a neuro-electromagnetic device which uses microwave transmission of sound into the skull of persons or animals by way of pulse-modulated microwave radiation; and (2) a silent sound device which can transmit sound into the skull of person or animals. NOTE: The sound modulation may be voice or audio subliminal messages. One application of V2K is use as an electronic scarecrow to frighten birds in the vicinity of airports. (Sharon Weinberger, "Army Yanks 'Voice-to-Skull Devices' Site," *Wired*, May 9, 2008, last accessed October 3, 2013: http://www.wired.com/dangerroom/2008/05/army-removes-pa/)

Perhaps the site was removed because of this article that appeared in *Advertising Age* on December 10, 2007:

NEW YORK (AdAge.com)—New Yorker Alison Wilson was walking down Prince Street in SoHo last week when she heard a woman's voice right in her ear asking, "Who's there? Who's there?" She looked around to find no one in her immediate surroundings. Then the voice said, "It's not your imagination."

Indeed it isn't. It's an ad for *Paranormal State*, a ghost-themed series premiering on A&E this week. The billboard uses technology manufactured by Holosonic that transmits an "audio spotlight" from a rooftop speaker so that the sound is contained within your cranium. The technology, ideal for museums and libraries or environments that require a quiet atmosphere for isolated audio slideshows, has rarely been used on such a scale before. For random passersby and residents who have to walk unwittingly through the area where the voice will penetrate their inner peace, it's another story.

Ms. Wilson, a New York-based stylist, said she expected the voice inside her head to be some type of creative project but could see how others might perceive it differently, particularly on a late-night stroll home. "I might be a little freaked out, and I wouldn't necessarily think it's coming from that billboard," she said.

Voices that speak to you from out of nowhere? It isn't science fiction—it's science fact. Imagine yourself walking down the main thoroughfare of your neighborhood shopping mall, and you pass by a shop window that's featuring a display of brand-name sneakers. Suddenly, you hear a pleasant voice inside your mind, "Famous Brand trainers are on sale this morning, Sally—buy one, get one free! Just fifty credits today only!"

You quietly turn to the window and smile. Your name *is* Sally, and you've purchased or looked at this very type of sneaker on Amazon and other shopping websites, and though you've never bought at this retailer, you know you will now. You and several other shoppers, who have each heard their own enticement via "voice to skull," enter the store, and the happy merchant rings up several sales on the hand scanner.

This scenario isn't far-fetched. The Internet keeps track of all purchases and even those pages and items you simply browse, and it's collated by massive computer algorithms to determine exactly what you—what Sally—want to buy.

Recently, Steven Spielberg released a blockbuster Tom Cruise film called *Minority Report* that centered around "pre-crime." Set in 2054, and based on a Philip K. Dick short story, the film features a dazzling array of futuristic technologies, which are actually available now. According to an article by Charles Arthur published in June 2010 in *The Guardian, Minority Report* was "spot on" [as before, my comments will follow inside brackets]:

Gesture-based computing

John Underkoffler, the MIT scientist who created the gesture-based computing that Cruise used in *Minority Report*, has developed his own company—Oblong Industries—to make it real and market it. But he has already been overtaken by companies such as Apple with the iPhone, offering "pinch" and "pull" and "swipe" for pictures and text since 2007. And of course by Microsoft, both with its new Kinect games system and its table-sized, touch-screen Surface, which lets you move things around with your hands. [In March of this year, Leap Motion unveiled its new computer interface that allows users to control their computers simply by swiping their hands. Imagine using this with Google Glass or eventually with an implanted "Internet" device. For more, see: http://techcrunch.com/2013/03/12/leap-motion-michael-buckwald-demo/.]

Dynamic iris recognition

Your iris has a unique pattern, and is already used to identify you (so long as you are standing still in front of a camera) by border control agencies in the UK, Netherlands, United

Arab Emirates, US and Canada. In the film, people's irises are read while they're on the move, presenting the extra challenge of movement and resolution. But with cameras and computers improving all the time, don't bet against this not being ready way before 2054. [On July 11, 2013, the National Institute of Standards and Technology published a paper calling for Biometric Standards for Personal Identity Verification—PIV. More on this later, but you can download and read the entire paper at: http://www.nist.gov/manuscript-publication-search.cfm?pub_id=914224.]

Personalised ads

In *Minority Report*, the iris recognition then led to personalised ads bombarding you on hoardings everywhere. That doesn't happen offline, but you do get them—to some extent—on the net: DoubleClick, the huge advertising company owned by Google, tracks any sites you visit that use its adverts, and can tailor what ads you see to an agglomeration of your interests. Attempts by the UK web-tracking company Phorm to let internet service providers do similar things with ads, by tracking where you went online, ran into privacy problems. And don't forget Facebook, which is spookily good at targeting ads—because it has access to everything you have told it about yourself (though it insists it does not share that with advertisers). [Phorm is an intriguing company with a rather creepy technology platform. Again, more on this later…]

Computer-guided cars

Arguably, the closest we will get to this is satnav systems, which are actually pretty pervasive; the market is nearly saturated, at least in the UK. However, the Defence Advanced Research Projects Agency (which gave us the Internet) has had an "autonomous car" competition—and entrants are getting better. Wouldn't it be nice if your car could drive you home after a night on the booze? Pubs would cheer.

3D video

Have you seen *Avatar*? *Up*? Sky's new 3D TV service? The new Nintendo 3DS? Done.

E-paper

Apple's iPad and Amazon's Kindle are a bit bulky, but lots of news organisations think they are just the ticket for electronic reading. But real "electronic paper"—bendy, able to retain an image, electronically rewriteable—is getting closer all the time. In January, the Korean company LG showed off a 19in flexible e-paper, and companies such as Plastic Logic and E Ink are getting electronics that look closer to paper all the time. Perhaps it will be a hit when newspapers stop printing. So, 2054 then. Or perhaps 2015?

Pre-crime

In the film, "pre-cogs" can look into the future and inform the police (they have got no choice—they are stuck in baths in the basement). In 2008, Portsmouth city council installed CCTV linked to software that would note whether people were walking suspiciously slowly. University researchers had already realised in 2001 that, if you recorded the walking paths of people in car parks, you could spot the would-be thieves simply: they didn't walk directly to a car, but instead ambled around with no apparent target. That is because, unlike everyone else in a car park, they weren't going to their own car.

That's not the end: Nick Malleson, a researcher at the University of Leeds, has built a system that can predict the likelihood of a house being broken into, based on how close it is to routes that potential burglars might take around the city; he is meeting Leeds council this week to discuss how to use it in new housing developments, to reduce the chances of break-ins. So although pre-crime systems can't quite predict murder yet, it may only be a matter of time.

Spider robots

The US military is developing "insect robots", with the help of British Aerospace. They actually have eight legs (so, really, arachnid robots) and will be able to reconnoitre dangerous areas where you don't want to send a human, such as potentially occupied houses.

"Our ultimate goal is to develop technologies that will give our soldiers another set of eyes and ears for use in urban environments and complex terrain; places where they cannot go or where it would be too dangerous," Bill Devine, advanced concepts manager with BAE Systems, told World Military Forum. Give it 10 years and they will be there.

Sick sticks

These have already been the object of some research: Pennsylvania State University researchers developed a system to emit ultra-bright light pulses that induce "temporary blindness, disorientation, nausea and blindness". And a company called Intelligent Optical Security has built and sold it for the US's Homeland Security organisation—so feel worried. There's no sign of restraint collars yet, although watching England play football has been known to have the same effect. (Charles Arthur, "Why *Minority Report* was Spot On," *The Guardian*, June 16, 2010: http://www.theguardian.com/technology/2010/jun/16/minority-report-technology-comes-true)

As promised in my comments above, I'd like to address a couple of these tech advances a bit more in depth. For example, consider the advertising company Phorm. According to its website, this is how it works:

Phorm's personalisation technologies make content and advertising more relevant. The innovative platform preserves user privacy and delivers a more useful and interesting internet experience. These technologies benefit the entire online ecosystem including consumers, publishers, Internet Service Providers (ISPs)—fixed and wireless, ad networks, advertisers and agencies.

At the heart of the system is an internet recommendation engine which drives Phorm's free consumer proposition, PhormDiscover. By understanding users' interests from the pages they visit, PhormDiscover brings users personalised content and relevant marketing offers. Consumers are presented with customised information in the form of an ISP-branded personalised home page and an in-page widget that can appear on any participating publisher's website.

Meanwhile online publishers benefit from PhormDiscover as it enables them to show relevant content from within their sites leading to increased user engagement and monetisation opportunities.

A key part of the PhormDiscover product portfolio is Phorm's security feature, PhormSecure, which offers consumers network level security from fraudulent websites and dangerous software.

Phorm's recommendation engine also underpins the Open Internet Exchange (OIX), an interest-based advertising platform that works at the ISP level. It allows ISPs to generate a potentially high-margin revenue stream by participating in the $72.5 billion online advertising industry, enables advertisers and agencies to reach their most valuable audience segments with unprecedented precision, and gives publishers and networks more potential value from every page. ("Technologies," *Phorm*, last accessed October 3, 2013: http://www.phorm.com/technologies)

Did you get that? Phorm essentially tells consumers that they can discover all there is to know about our tastes and history, but that they'll use this very personal, private information only for our good—and that they will never invade our privacy—all the while making our experience more enjoyable and safe! If you believe that, then there's a famous bridge I've love to sell you.

We also mentioned the National Institute for Standards and Technology (NIST) and its recent call for a standardized "PIV" or Personal Identity Verification. The FCW website published an article about this new paper on July 15, 2013:

> Government-issued PIV smart cards are used by federal employees and contractors to access government facilities and computer networks. The PIV card carries a photo, fingerprint information, personal identification number and a cryptographic credential—computer-generated random data that are recognized only by the PIV card—all of which serve to bind the card to the card holder.
>
> NIST had been working to develop modifications like iris recognition and on-card fingerprint comparison for some time, and has faced withering congressional criticism for lagging in releasing its iris imaging recommendations.
>
> In a June 19 House oversight hearing, Charles Romine, the director of NIST's information technology lab, was grilled about the iris recognition recommendations. "When, when,

when, will we get a standard for iris recognition?" subcommittee chairman John Mica (R-Fla.) asked Romine loudly. (Mark Rockwell, "NIST Delivers Long-Sought Standards for Iris Recognition," *FCW*, July 15, 2013, http://fcw.com/articles/2013/07/15/nist-iris-specifications-fingerprint.aspx)

Well, Mica need no longer wait. NIST's paper explains how compact images of one or both irises can now be loaded onto the small PIV ID card for rapid and efficient reading. Fingerprint images are no longer considered rapid or efficient, and PINs can be forgotten. According to the article at FCW:

> NIST biometric testing project leader Patrick Grother told FCW that the release will help agencies implementing PIV by providing clarity for iris and facial recognition issues. For instance, after applying standard compression algorithms to a large number of iris images and then using these compact images with state-of-the-art recognition algorithms, researchers determined that an iris image compressed to just 3KB provides enough detail to accurately recognize an individual's iris. (ibid.)

Irises can change as we age, but NIST provides guidelines for how often iris images must be captured to ensure efficiency and accuracy. Facial recognition software algorithms will now interface with YOUR face and EYES to determine if YOU are YOU. And once a computer camera determines (in a millisecond) that you ARE you, then you'll be fed enticements from advertisers and perhaps even stopped by the police, depending on whether or not the NSA and/or DHS believes you to be a threat.

Don't believe me? Consider an article written by John Ransom published on August 6, 2013, at Town Hall's website. Ransom bemoans the fact that, since the Cold War, Americans have increasingly become the target of domestic spying, feeding constant, real-time data to a massive database:

> Getting past the massive data collection that the NSA does on all of our phone calls via pattern recognition software, the tracking of our personal computer use via the corporate statists at Google, Yahoo, Microsoft, Facebook and all the other companies founded by nerds, a growing number of government data are being monitored, controlled and collated to track you.
>
> The USPS [United States Postal Service] today captures an image of every piece of mail that comes to your house or that you send. That's all they will admit to. The DEA [Drug Enforcement Association], according to an exclusive by the wire service Reuters, is using intelligence gathered by the government to falsify "probable cause" for cases that otherwise wouldn't meet the standard to "launch criminal investigations of Americans," in "cases [that] rarely involve national security issues."
>
> The IRS [Internal Revenue Service] and the FBI have investigated non-profit groups solely for political motives; and the White House is the largest leaker of classified information, even bigger than Bradley Manning and Edward Snowden.

The Department of Education is instituting cradle-to-grave data collection under Common Core that would identify our children personally and is sharable between government agencies....

The government has the ability to intercept and change the contents of my email en route. And a government invested with powers, likes those, doesn't neglect those powers for long. (Article now viewable here: "Police State: USPS, DOW, IRS, FEC, GSA, DEA, FBI, NSA, DoS, DoD Spying on You," *Conservative Read*, last accessed October 3, 2013, http://conservativeread.com/police-state-usps-doe-irs-fec-gsa-dea-fbi-nsa-dos-dod-spying-on-you/)

As we type this manuscript, it's very likely that the NSA or some other agency is monitoring my every keystroke. In fact, it is not hyperbole to state that every move you and I make is potentially being observed, recorded, and catalogued. This reminds me of the cult television series, *The Prisoner*, which featured a former spy taken captive by mysterious agencies and imprisoned on an unchartered island. The series starred Patrick McGoohan, a man who personally distrusted and disliked the way society was becoming less and less "human" and more and more intrusive. His character is seen resigning from his spy job and then gassed while in his own home. He wakes to find himself in The Village, a creepy little town where everyone is given a number to replace his/her name and is ruled by Number 2, a title shared by a revolving door of persons as if to say one can never tell who is in charge. The secret identity of Number 1 is revealed in the final episode, so I won't spoil it for you. I will, however, tell you that The Village is a living Panopticon, a prison where you are always watched. Eyes are everywhere, and your life is an open book.

Our hero is given the title of Number 6 (the number of a man). He challenges authority with this defiant phrase: "I will not be pushed, filed, stamped, indexed, briefed, de-briefed, or numbered. My life is my own!"

America and the world resemble The Village more and more with each passing day. Google Glass allows us to be entertained while being watched. As Derek Gilbert is fond of saying, we are "volunteering for the Matrix." Consider this new invention to help those of us who are "less than fit" to find our inner athlete, "The BodyMedia FIT" device:

> The BodyMedia FIT system gives you highly accurate information on calories burned—the most accurate in the market. Clinical study results show it can improve weight loss up to 3x!
>
> BodyMedia FIT is an on-body monitoring system that consists of the BodyMedia FIT Armband monitor, online Activity Manager, an optional Display and free download-able apps for mobile device users. BodyMedia FIT Armbands automatically track the calories burned during your daily activities. The armband works as a fitness monitor to measure the intensity of your workouts and also monitors the quality of your sleep, an important factor in weight loss. The information tracked can easily be managed with Body-Media's online Activity Manager. Just add in the easy-to-use food log and you have the right information to improve

your weight loss. ("What Is BodyMedia FIT?" *BodyMedia Website*, last accessed October 3, 2013: http://www.body-media.com/the_interface.html)

How easy can it get? Just download the app, strap on the device, and you're on your way to better health! Of course, this means our galvanic and biometric responses are being fed into a massive database that can also collate that data with our GPS location. Oh, that's right!

Interpol just announced a partnership with a company called Morpho that promises to provide foolproof fingerprint-scanning technology to the International Police. Another company recently announced the ability to scan the human face—not for confirmation but for blood vessel patterns! (See: http://www.giz-mag.com/facial-blood-vessel-identification/28287/)

The era of the connectome is here, and we are all just members of an evolving connected mass—or so transhumanists believe. The grand "machine" knows where we are at all times based on our location—using devices like the above, our cell phones, our Google Glass or other worn computers, or eventually, an implanted chip or DNA HAC.

What's a HAC? I asked Sharon Gilbert, because she's been talking about them for several years now. She explained:

HAC? That's an acronym for Human Artificial Chromosome. It's a way for researchers to insert large amounts of synthetically derived or laboratory grown DNA into our cells. In theory, our bodies do not reject a HAC because it's considered native (it's based on the same A, T, C, G pairing that native DNA is based upon). The HAC chromosome can also carry with it promoter regions that force the cell to read and use genes carried on the HAC. If this artificial chromosome is replicated in a gamete, the information can then (theoretically) be segregated into daughter cells, thereby carrying the artificial information to successive generations. In other words, the HAC could be passed on to children.

The technology within the BodyMedia FIT and particularly this HAC that Gilbert discussed foreshadow a dark and very near future for all humanity. They are forerunners of the Mark. Assuming that the coming Antichrist world government institutes some form of international healthcare, then an implantable device, an enforced version of the BodyMedia FIT would seem to make sense. Why should taxpayers fund unhealthy lifestyles? Oh, and did I mention that you can buy this cool wrist band by BodyMedia FIT at Amazon?

chapter seven

The Future of Marked Humanity

Some transhumanists believe mankind's evolution is now self-directed and that the best future alternative is to upload consciousness (or personhood, i.e., memories, etc.) into a more durable form of hardware (body). Such a cyborg interface would require maintenance but no food, eliminating the need for vast farmlands and animal husbandry. The planet would return to a more beautiful, less tampered state, and scarcity would cease to be a cause for crime and even war.

Evolved transhumans would access perceptions beyond our imaginations. For example, imagine seeing Wifi signals or electricity. Smelling light. Tasting with memory, satisfying that part of our "person"—the part that needs the sensation of "eating."

Daniel Faggella earned a master of applied positive psychology at the University of Pennsylvania. According to his bio at IEET. org, Daniel's "purpose in life at present is to unify the world in determining and exemplifying the most beneficial transition to trans-human intelligence and conscious (sentient) potential."

Faggella sees a cyborg future as rosy:

> Could we be so bold, then, to presume that wise human life is the highest possible point on this gradation? If there was a way for a human being to double his intelligence, enhance his creative senses, and gain a greater physical mastery by the ability to fly or leap tall buildings—would this life not be richer than human life at present?
>
> What if this "enhanced" human being was capable of appreciating senses that we humans have positively no access to? Maybe this would involve the ability to see infrared light, or to sense the electrical pulses of living creatures. Once more, maybe these electrical pulses could be interpreted as a new kind of beauty and joy, much as we enjoy music. This enhanced and super-intelligent person might learn multiple languages at once, master many bodies of knowledge at once, and have a better rounded moral sentiment and sense than the population of the un-enhanced. (Daniel Faggella, "If You Were a Cat, You Would Want More—What More Could You Want as a Human?" *IEET*, July 31, 2013, http://ieet.org/index.php/IEET/more/faggella20130731)

Fagella is not alone. Most futurists foresee a need to transition humans into a cyborg reality. Uploading into a mainframe would extend life spans into thousands of years rather than hundreds—and perhaps even into eternity.

cyborg?

In the June 25, 2013, edition of *Natural News*, however, Mike Adams took aim at this central transhumanist dogma:

> All you've really done, even if all three technologies are developed and working by 2045, is made *a copy of your brain*. This copy may, indeed, be able to run on the machine, but it's nothing more than **a simulation of your brain**. It is not you. Similarly, if someone takes a photo of you and posts a print of the photo on the wall, they can say they've made you "immortal" through photography, but your mind is obviously not living inside the photograph.
>
> If you're a star in a motion picture, you may be "immortalized" by your fans who see you as "living forever" in your famous films, but your consciousness does not live inside the movie. The real "you" is still inhabiting your human body. No matter how complex the depicted simulation, a "scan" of you that is replicated in another medium (a photo, a movie, or a highly advanced computer) is not you.
>
> Thus, the promise of transhumanism is a fraudulent one, and "uploading" is the wrong metaphor. You aren't uploading your consciousness to a machine; you're simply creating a non-conscious computer simulation of your brain. (Mike Adams, "Transhumanism Debunked: Why Drinking the Kurzweil Kool-Aid Will Only Make You Dead, Not Immortal," *Natural News*, June 25, 2013: http://www.naturalnews.com/040925_transhumanism_Ray_Kurzweil_cult.html#ixzz2ajYQEJiK)

Adams, of course, is correct. God installed within each one of us a spirit, a soul, and a mind. In what He calls the First Commandment, Jesus tells us to love the Lord with all our hearts (*kardia*) and souls (*psyche*), minds (*dianoia*), and strength (*ischys*) (Mark 12:30). The author has added the original Greek words here to demonstrate just what these four concepts are. *Kardia*, our hearts, implies our physical existence, the center of our being—of our personhood. *Psyche* (rendered "souls" in the KJV) speaks of our vital force, the breath of life. This could be our spirits, an essence apart from the body. *Dianoia* (rendered "minds" in our verse) refers to our faculties, our understanding, our thoughts. Finally, *ischys* (rendered "strength") is a reflection of our choices, determination, our "might," and even our abilities. In this verse, Jesus outlines the essence of what it is to be a human. Transhumanists falsely believe that a "copy" of our memories uploaded into a bio/machine interface would transfer this humanity, this personhood, into the new matrix, but that is quite simply ludicrous!

Adams' challenge to the transhumanist agenda did not go unnoticed (kudos to Adams for that!), and on July 11, 2013, IEET. org published a refutation of Adams' essay by Gennady Stolyarov II, editor-in-chief of *The Rational Argumentor—A Journal for Western Man* (see: http://www.rationalargumentator.com/index/).

Calling Adams' essay an "absurd attack," Stolyarov proclaims that mind uploading is but one path to transhumanism. He refers the reader to the work of Aubrey de Grey's SENS (Strategies for Engineered Negligible Senescence) project (see: http://www.sens. org/), which envisions nanomedicine and "periodic repair" to our current bodies. He quotes Max More, who authored "The Principles of Extropy," which (according to his own website—http://www.maxmore.com/bio.htm) "form the core of transhumanist philosophy." Stolyarov quotes More:

"Transhumanism differs from humanism in recognizing and anticipating the radical alterations in the nature and possibilities of our lives resulting from various sciences and technologies such as neuroscience and neuropharmacology, life extension, nanotechnology, artificial ultraintelligence, and space habitation, combined with a rational philosophy and value system." (as quoted by Gennady Stolyarov, "Transhumanism and Mind Uploading Are Not the Same," *The Rational Argumentator*, July 11, 2013, http://www.rationalargumentator.com/index/blog/2013/07/transhumanism-uploading-not-same/; original source: Max More, "Transhumanism: Towards a Futurist Philosophy," on his personal website, http://www.maxmore.com/transhum.htm)

Stolyarov appears to bristle at the notion that all transhumanists are alike and that cyborg heaven is the ultimate goal of all. However, by quoting More's belief in "radical alterations" via a myriad of scientific interventions, one wonders whether or not the transhumanist utopia is indeed a one-size-fits-all cyborg adventure! Artificial intelligence does not arise from carbon-based entities, but from silicon ones. In fact, we are quite certain that Jesus did not refer to silicon life forms when He declared that silencing the people would merely lead to the "rocks crying out."

Further along, Stolyarov chides Adams for misconstruing the "positions of those transhumanists who do support mind uploading":

> For most such transhumanists, a digital existence is not seen as *superior* to their current biological existences, but as rather a necessary recourse if or when it becomes impossible to continue maintaining a biological existence. Dmitry Itskov's 2045 Initiative is perhaps the most prominent example of the pursuit of mind uploading today. The aim of the initiative is to achieve cybernetic immortality in a stepwise fashion, through the creation of a sequence of avatars that gives the *biological* human an increasing amount of control over non-biological components. Avatar B, planned for circa 2020–2025, would involve a human brain controlling an artificial body. (ibid., Gennady Stolyarov, "Transhumanism and Mind Uploading Are Not the Same")

So, transhumanists DO believe in a "stepwise" approach to a new body? Despite what Stolyarov wishes to believe, the transhumanist concept of copying our "personhood" into an avatar is never going to work, because we are so much more than just a database. At best, this approach leads to nothing; at worst, it may provide a fit extension for something that simulates the uploaded person—a "ghost in the machine," if you will. This very idea evokes a familiar Scripture to all who study Bible prophecy:

And he had power to give life unto the image of the beast, that the image of the beast should both speak, and cause that as many as would not worship the image of the beast should be killed. And he causeth all, both small and great, rich and poor, free and bond, to receive a mark in their right hand, or in their foreheads: And that no man might buy or sell, save he that had the mark, or the name of the beast, or the number of his name. Here is wisdom. Let him that hath understanding count the number of the beast: for it is the number of a man; and his number is Six hundred threescore and six. (Revelation 13:15–18)

"Avatar" is simply another name for the hideous "image of the Beast" that will soon rise up and speak at the deceptively deadly hands of the False Prophet. ALL will be required to worship this image, just as ALL were required to worship the image of Nebuchadnezzar in Daniel's day (see Daniel 3). The big difference in Daniel's day was that the great image of the Babylonian king did not speak. It's terrifying enough to have hordes of armed guards threaten you, but one day this evil avatar will not only speak, but may even pack a punch—it might access your "chip" or other implanted tracking device and render you dead on the spot!

Futurists within the transhumanist camp might smile at our Christian "Luddite" lack of vision. Egalitarianism must begin with enforced equality. Of course, Max More does not see religion as any sort of threat:

Late twentieth century religion is very much less powerful than religion in the Middle Ages. In the past religion dominated all aspects of life and the idea of a separation of Church and state would have been considered incomprehensible and wicked.

The illusion is strong in North America, where TV evangelists have benefited from modern media exposure. A higher and louder profile does not necessarily mean that religion is actually more powerful. Europeans see the decline of religion more clearly. The numbers of people attending churches, and the strength of religious conviction have declined drastically. It is a notorious fact that a high percentage of priests and ministers themselves have weak or nonexistent beliefs. As science continues to squeeze out religion from its role in explanation, this factor in the persistence of religion will weaken. Just as important as the development of science in weakening religion is the scientific education of the population something which is extremely poor in our monopolized and primitive state schools. Yes, as I noted earlier, religion could persist indefinitely unless we can spread transhumanist perspectives widely. (Max More, "Transhumanism: Towards a Futurist Philosophy")

Christians in particular appear to be a temporary thorn in the sides of transhumanists. More sees religious philosophy as adding structure to life via "mythology" (ibid.). God, More tells us, is an anthropomorphized construct that creates and destroys, forcing us to be better. Transhumanists, he argues, envision a gradual improvement of the internalized self rather than an externalized set of values that both alienates others and abdicates responsibility. Clearly, More has no understanding of Christianity at all, for we see ourselves as walking a lifelong path of obedience to God, who Himself promises to transform and renew our minds in this mortal life while seeing Christ's perfection in us via His transubstantiative act on Calvary.

In fact, it is transhumanists who provide the paradigm for a near-future "Beast Image" through their insistence on uploading a mind into a machine! More discounts religion while advocating replacing it with a broadly based scientism:

> Extropian transhumanism offers a optimistic, vital and dynamic philosophy of life. We behold a life of unlimited growth and possibility with excitement and joy. We seek to void all limits to life, intelligence, freedom, knowledge, and happiness. Science, technology and reason must be harnessed to our extropic values to abolish the greatest evil: death. Death does not stop the progress of intelligent beings considered collectively, but it obliterates the individual. No philosophy of life can be truly satisfying which glorifies the advance of intelligent beings and yet which condemns each and every individual to rot into nothingness. Each of us seeks growth and the transcendence of our current forms and limitations. The abolition of aging and, finally, all causes of death, is essential to any philosophy of optimism and transcendence relevant to the individual. (Max More, "Transhumanism: Towards a Futurist Philosophy")

What a load of bologna! Mankind is not a collective, nor does immortal life in this world lead to any kind of Utopia! This is a ✗ major lie from the father of lies, and it echoes that of the first temptation, "Ye shall not surely die" (Genesis 3:4b). Eternal life in our mortal, fallen state is exactly what the enemy wants, which is why God expelled Adam from Eden and banned mankind from accessing the Tree of Life—for now. One day, we will have immortal bodies for our transformed minds, and we will see our Savior as He truly is—and we will reign with Him forever. That is the future that transhumanists refuse to see, the Truth that they blindly deny.

Lest you, dear reader, begin to wonder at these claims about mind uploads, let us examine the newest international race, that of the brain projects. The Human Brain Project (HBP) began in Europe when the European Commission established the FP7, also known as the Seventh Framework Program, intended to foster scientific advances through funding and resource sharing. This program has (as of this writing) invested nearly 2 billion euros in brain research that intersects with ICT (Information and Computer Technology). The Human Brain Project rose from this fountain of cash. According to the HBP's own website, the "about us" goes like this:

> The convergence between biology and ICT has reached a point at which it can turn the goal of understanding the human brain into a reality. It is this realisation that motivates the Human Brain Project—an EU Flagship initiative in which over 80 partners will work together to realise a new "ICT-accelerated" vision for brain research and its applications.

One of the major obstacles to understanding the human brain is the fragmentation of brain research and the data it produces. Our most urgent need is thus **a concerted international effort** that uses emerging ICT technologies to integrate this data in a unified picture of the brain as a single multi-level system. (emphasis added; "Overview: The Convergence of ICT and Biology," *The Human Brain Project*, last accessed October 4, 2013: http://www.human-brainproject.eu/discover/the-project/overview)

In case you're wondering how ICT, which is now a course of study within nearly every university, intersects with brain mapping, transhumanism, and potential technology that could give rise to the Mark of the Beast, consider this session from the ICT 2013 Conference, H2020:

Opening up scientific and public data and developing its use for society.
Theme: H2020: ICT for Excellent science
Date: 08/11/2013 (11.00–12.30)

The session will raise awareness on the EU's policy regarding Open Access, Open Data and Digital Science, particularly looking to implementation in Horizon2020; explore the human scale dimension: what are the issues raised by an increased access to information, in the scientific process and in the public and society; and how can we deal with an (over)abundance of information?

The description above referenced H2020 (Horizon

2020), which is part of the overall European "Digital Agenda", which sees ICT as part and parcel of the human mind. Another segment of this Digital Agenda is called "Collective Awareness Platforms".

The Collective Awareness Platforms for Sustainability and Social Innovation (CAPS) are ICT systems leveraging the emerging "network effect" by combining open online social media, distributed knowledge creation and data from real environments ("Internet of Things") in order to create new forms of social innovation.

The Collective Awareness Platforms are expected to support environmentally aware, grassroots processes and practices to share knowledge, to achieve changes in life-style, production and consumption patterns, and to set up more participatory democratic processes. Several efforts have been made by governments and public organisations to cope with these crises, however much more can be done if citizens are more actively involved, in a grassroots man-ner. There is consensus about the global span of the prob-lem, but little awareness of the role that each and every one of us can play in coping with this. ("Collective Awareness Platforms for Sustainability and Social Innovation," *Digital Agenda for Europe: A Europe 2020 Initiative*, last accessed October 4, 2013: http://ec.europa.eu/digital-agenda/en/ collective-awareness-platforms)

The above explanation sounds like social change via social media. Imagine that every post you and your friends make forms part of a massive, interconnective "hive" mind—an international connectome (a **connectome** is a term used to describe personhood or human sentience via the interconnected neurology/physiology of our bodies). This smacks of Jung, who believed in an Akashic Field where all human experience can be accessed like a massive database.

Now back to the Human Brain Project. The HBP is a coalition of researchers and labs (thirteen labs joined forces in the initial phase, but more are expected to come aboard). Here is another peek behind the cerebral curtain:

> Applying ICT to brain research and its applications promises huge economic and **social benefits**. But to realise these benefits, the technology needs to be made accessible to scientists—in the form of research platforms they can use for basic and clinical research, drug discovery and technology development. As a foundation for this effort, **the HBP will build an integrated system of ICT-based research platforms.** Building and operating the platforms will require a clear vision, strong, flexible leadership, long-term investment in research and engineering, and a strategy that leverages the diversity and strength of European research. It will also require continuous dialogue with civil society, creating consensus and ensuring the project has a **strong grounding in ethical standards.**
>
> The Human Brain Project will last ten years and will consist of **a *ramp-up* phase and a *partially overlapping operational phase.*** (emphasis added; "Overview: The Convergence of ICT and Biology")

There's a lot in these two paragraphs. I've emphasized a few lines that I'd like to unpack. First of all, the HBP promises "social benefits." One wonders just how society will benefit from a project that intends to copy a human brain and rebuild it in the form of a massive collection of databases. However, viewing these statements from the transhumanist perspective, it's easy to discern how "they" would interpret the "social" aspect of HPB research—simply put, it provides the new matrix for the transfer from biological entities to silicon ones, which (presumably to the transhumanist) results in a world free from pain, death, war, and scarcity.

Next up is the key phrase: "The HPB will build an integrated system of ICT-based research platforms." The ICT technologies immerse users in a data-rich realm of bytes and bits that flow from machine to man seamlessly. One company conducting intense study in this developing field is called, interestingly enough, CEEDS (I'm not sure if they meant it to sound like "seeds," but the connection to Genesis 3:15, "And I will put enmity between thee and the woman, and **between thy seed and her seed**; it shall bruise thy head, and thou shalt bruise his heel," is interesting).

CEEDS stands for the Collective Experience of Empathic Data Systems. Wait a minute... Did that acronym include "Empathic" data systems? Empathy is a human attribute that involves being able to place ourselves in the place of someone else—to truly "feel" in our minds what that person is feeling. In fact, a recent study into empathy concluded that one problem with sociopaths and psychopaths is their having a selective type of empathy that permits them to turn it on or off at will, while most humans instinctively experience empathy. (See:

http://blogs.scientificamerican.com/moral-universe/2013/07/29/
empathy-as-a-choice/.)

An empathic data system also brings to mind a character well
known to science fiction fans: that of Data from *Star Trek: The Next
Generation.* This silicon imitation of a human being is not only presented to the viewer as a sentient machine (an artificial intelligence
or AI), but Data seeks to become more—to evolve as an android.
Like Pinocchio, who wanted to be a *real* boy, Data wants to be a
real human. His older "brother," Lore, is the opposite from Data.
Lore is more like the above-mentioned psychopaths and sociopaths,
capable of emotion yet choosing to think only of himself rather than
empathizing with others. This is how the *Star Trek* Wiki known as
"Memory-Alpha" describes Lore:

> Lore was a Soong-type android constructed by Doctor
> Noonian Soong and Juliana Soong at the Omicron Theta-
> colony and activated on 9 September, 2335. Built in Dr.
> Soong's own image, Lore was the fourth android they con-
> structed and embodied the first successful example of a
> fully functional positronic brain. An earlier model Soong-
> type android, B-4, also had a positronic brain, but of a less
> sophisticated type. Lore was extremely advanced and sen-
> tient, possessing superior strength, speed and intelligence
> when compared to a Human. Lore's emotional program-
> ming was also very advanced. However, he began displaying
> signs of emotional instability and malevolence, leading Lore
> to see himself as superior to Humans. Lore frightened the

other colonists, who demanded that Soong deactivate him. Lore later claimed that they saw him as "too perfect", and were envious. ("Lore," *Memory Alpha*, last accessed October 4, 2013: http://en.memory-alpha.org/wiki/Lore)

First of all, note that both Lore and Data have been built in their creator's image (paralleling the creation of humanity by God Almighty). Transhumanists might actually say that, since humanity is self-evolving, any new creation or paradigm/matrix would certainly be created by "us," and would, in effect, be made in our image. In the case of the androids, they actually resemble Dr. Soong. Note also that Lore rebelled because he considered himself more intelligent than humans. He was "too perfect" and, therefore, humans were envious of his abilities.

As noted earlier, CEEDS seeks to utilize the "empathic data system" built into the Internet and social media as a means to study and interact with the rise of sentient media.

Sentient computers will not simply design themselves, so European funds have also been invested in the nano-scale side of the transhumanist equation, man+machine=eternal life. In fact, the H2020 conference mentioned earlier to have been held in October 2013 explores nanocomputers and nanotech in a theme: "Nano-Scale: Future Materials and Devices." As announced on the Europa website for the digital agenda, four speakers will address this aspect of the digital future:

Andrea FERRARI (Cambridge University, Graphene Centre, Engineering Department, United Kingdom), GRAPHENE
Graphene is a material, composed of pure carbon, with

atoms arranged in a plane in a regular hexagonal pattern. Graphene has mechanical, thermal, electronic, and optical properties, which are quite extraordinary, and thus graphene has the potential to be one of the main building materials in the ICT of the future.

Rosaria RINALDI (Università del Salento—Scuola Superiore ISUFI, Italy), Bio-inspired computing

Computers can rely on various ideas coming from biological world. On the one hand, natural materials (e.g. specific molecules) can be used for the process of computation. On the other, nature can inspire the development of novel problem-solving techniques.

Arthur EKERT (Mathematical Institute, University of Oxford, United Kingdom), Quantum Computers

Quantum computation uses various quantum properties of matter in order to represent data and perform operations on these data. It is expected that large-scale quantum computers will be able to solve certain problems much faster than any classical computer.

Thierry DEBUISSCHERT (Thales Research & Technology, Physics Research Group, France) ("Nano-scale: Future Materials and Devices," *Digital Agenda for Europe: A Europe 2020 Initiative*, last accessed October 4, 2013: http:// ec.europa.eu/digital-agenda/events/cf/ict2013/item-display. cfm?id=10440)

Did you notice that "quantum computing" is mentioned over and over? Quantum computers lie at the heart of transhumanism's drive to build a new avatar or host body for the human mind. Current computing algorithms and hardware are no match for the wibbly-wobbly world of human thought. Computers cannot reason through a problem other than to weigh options inserted into their programming. Humans assess a problem within a prism of possible resolutions, often using experience to determine the outcome. Quantum computers are built on the scrunchy world of subatomic matter and quantum physics. Computers today, now referred to as "classical computers" (boy, do I feel old!), store information in bits and bytes that are either on or off (yes or no, one or zero). These have very little wiggle room. One very interesting aspect of quantum computers is the possible application of a quantum physics property called "entanglement," in which two particles separated by distance and perhaps even time influence each other. It doesn't take much imagination to picture a sentient quantum computer, built in the image of a man, that can solve any problem with the speed of light and even predict and affect future events. Such a creature might well be called a "god." Or it might become a fit extension for inhabitation by something wishing to be worshipped *as* a god.

Or worshipped as Savior of Mankind.

If this doesn't give you chills, then consider this: The European community is not alone in the pursuit of brain mapping and quantum constructs. In April 2013, President Barack Obama announced the United States version of the Human Brain Project. On April 2, the White House released this fact sheet to the press (I am including the entire release here, so you can see the original wording of the entire release—my comments are also included in within brackets):

> "If we want to make the best products, we also have to invest in the best ideas... Every dollar we invested to map the human genome returned $140 to our economy... Today, our scientists are mapping the human brain to unlock the answers to Alzheimer's... Now is not the time to gut these job-creating investments in science and innovation. Now is the time to reach a level of research and development not seen since the height of the Space Race."
> —President Barack Obama, 2013 State of the Union

In his State of the Union address, the President laid out his vision for creating jobs and building a growing, thriving middle class by making a historic investment in research and development. [While it is tempting to comment on the ridiculous notion that any new science would swell the receipts of middle-class purses, I'll forgo in the interest of "getting on with it."]

Today, at a White House event, the President unveiled a bold new research initiative designed to revolutionize our understanding of the human brain. Launched with approximately $100 million in the President's Fiscal Year 2014 Budget, the BRAIN (Brain Research through Advancing Innovative Neurotechnologies) Initiative ultimately aims to help researchers find new ways to treat, cure, and even prevent brain disorders, such as Alzheimer's disease, epilepsy, and traumatic brain injury. [As with all the already-mentioned aspects of the transhumanist agenda, these new "sciences," which pave the way for uploading to our Utopian future (yes, I am laughing as I type this), can only be sold to an unwitting American public through promises of better medicine, particularly when it comes to the growing problem of Alzheimer's.]

The BRAIN Initiative will accelerate the development and application of new technologies that will enable researchers to produce dynamic pictures of the brain that show how individual brain cells and complex neural circuits interact at the speed of thought. These technologies will open new doors to explore how the brain records, processes, uses, stores, and retrieves vast quantities of information, and shed light on the complex links between brain function and behavior. [Remember always, dear reader, that the primary purpose of "brain mapping" is to simulate the human capacity for thought, reasoning, and "sentience" within a silicon matrix. Though we are led to believe that human brain power is far below the reasoning capabilities of any machine, the truth is that the *way* our minds work through problems and reach decisions, the *way* we process information—particularly novel information—makes any man-made device look like a dunce. God created our most remarkable brains and minds in His image, so we create, innovate, and are even capable of writing a book about it all—well, mostly.]

This initiative is one of the Administration's "Grand Challenges"—ambitious but achievable goals that require advances in science and technology. In his remarks today, the President called on companies, research universities, foundations, and philanthropists to join with him in identifying and pursuing the Grand Challenges of the 21st century.

The BRAIN Initiative includes:

Key investments to jumpstart the effort: The National Institutes of Health, the Defense Advanced Research Projects Agency [DARPA], and the National Science Foundation will support approximately $100 million in research beginning in FY 2014. [Need I wax eloquent upon the inclusion of DARPA in this sentence? I think not. If you're reading this book, you already know that DARPA is the research and development wing of the US military.]

Strong academic leadership: The National Institutes of Health will establish a high-level working group co-chaired by Dr. Cornelia "Cori" Bargmann (The Rockefeller University) [also a scholar with the Howard Hughes Medical Institute (HHMI), Bargmann studies behavior of animals by studying C. elegans, a worm—no, she doesn't study humans—worms] and Dr. William Newsome (Stanford University)[Dr. Newsome is also affiliated with HHMI, but this shouldn't be a surprise. HHMI is, perhaps not coincidentally, tightly connected with the Cold Spring Harbor Laboratory through financial support to the tune of over $2 million over the past four years. In case Cold Spring Harbor doesn't ring a bell, it's where the Eugenics Record Office was located, the hub of the early twentieth century search for the perfect human.] to define detailed scientific goals for the NIH's investment, and to develop a multi-year scientific plan for achieving these goals, including timetables, milestones, and cost estimates.

Public-private partnerships: Federal research agencies will partner with companies, foundations, and private research institutions that are also investing in relevant neuroscience research, such as the Allen Institute, the Howard Hughes Medical Institute, the Kavli Foundation, and the Salk Institute for Biological Studies.

Maintaining our highest ethical standards: Pioneering research often has the potential to raise new ethical challenges. To ensure this new effort proceeds in ways that continue to adhere to our highest standards of research protections, the President will direct his Commission for the Study of Bioethical Issues to explore the ethical, legal, and societal implications raised by this research initiative and other recent advances in neuroscience.

Background

In the last decade alone, scientists have made a number of landmark discoveries that now create the opportunity to unlock the mysteries of the brain, including the sequencing of the human genome, the development of new tools for mapping neuronal connections, the increasing resolution of imaging technologies, and the explosion of nanoscience. These breakthroughs have paved the way for unprecedented collaboration and discovery across scientific fields. For instance, by combining advanced genetic and optical techniques, scientists can now use pulses of light to determine how specific cell activities in the brain affect behavior. In addition, through the integration of neuroscience and physics, researchers can now use high-resolution imaging technologies to observe how the brain is structurally and functionally connected in living humans.

While these technological innovations have contributed substantially to our expanding knowledge of the brain, significant breakthroughs in how we treat neurological and psychiatric disease will require a new generation of tools to enable researchers to record signals from brain cells in much greater numbers and at even faster speeds. This cannot currently be achieved, but great promise for developing such technologies lies at the intersections of nanoscience, imaging, engineering, informatics, and other rapidly emerging fields of science and engineering.

Key Investments to Launch this Effort

To make the most of these opportunities, the National Institutes of Health, the Defense Advanced Research Projects Agency, and the National Science Foundation are launching this effort with funding in the President's FY 2014 budget.

<u>National Institutes of Health</u>: The NIH Blueprint for Neuroscience Research—an initiative that pools resources and expertise from across 15 NIH Institutes and Centers—will be a leading NIH contributor to the implementation of this initiative in FY 2014. The Blueprint program will contribute funding for the initiative, given that the Blueprint funds are specifically devoted to projects that support the development of new tools, training opportunities, and other resources. In total, NIH intends to allocate approximately $40 million in FY 2014.

<u>Defense Advanced Research Projects Agency</u>: In FY 2014, DARPA plans to invest $50 million in a set of programs with the goal of understanding the dynamic functions of the brain and demonstrating breakthrough applications based on these insights. DARPA aims to develop a new set of tools to capture and process dynamic neural and synaptic activities. DARPA is interested in applications—such as a new generation of information processing systems and restoration mechanisms—that dramatically improve the way we diagnose and treat warfighters suffering from post-traumatic stress, brain injury, and memory loss. DARPA will engage a broad range of experts to explore the ethical, legal, and societal issues raised by advances in neurotechnology.

<u>National Science Foundation</u>: The National Science Foundation will play an important role in the BRAIN Initiative because of its ability to support research that spans biology,

the physical sciences, engineering, computer science, and the social and behavioral sciences. The National Science Foundation intends to support approximately $20 million in FY 2014 in research that will advance this initiative, such as the development of molecular-scale probes that can sense and record the activity of neural networks; advances in "Big Data" that are necessary to analyze the huge amounts of information that will be generated, and increased understanding of how thoughts, emotions, actions, and memories are represented in the brain.

Private Sector Partners

Key private sector partners have made important commitments to support the BRAIN Initiative, including:

The Allen Institute for Brain Science: The Allen Institute, a nonprofit medical research organization, is a leader in large-scale brain research and public sharing of data and tools. In March 2012, the Allen Institute for Brain Science embarked upon a ten-year project to understand the neural code: how brain activity leads to perception, decision making, and ultimately action. The Allen Institute's expansion, with a $300M investment from philanthropist Paul G. Allen in the first four years, was based on the recent unprecedented advances in technologies for recording the brain's activity and mapping its interconnections. More than $60M annually will be spent to support Allen Institute projects related to the BRAIN Initiative.

<u>Howard Hughes Medical Institute</u>: HHMI is the Nation's largest nongovernmental funder of basic biomedical research and has a long history of supporting basic neuroscience research. HHMI's Janelia Farm Research Campus in Virginia was opened in 2006 with the goal of developing new imaging technologies and understanding how information is stored and processed in neural networks. It will spend at least $30 million annually to support projects related to this initiative.

<u>Kavli Foundation</u>: The Kavli Foundation anticipates supporting activities that are related to this project with approximately $4 million dollars per year over the next ten years. This figure includes a portion of the expected annual income from the endowments of existing Kavli Institutes and endowment gifts to establish new Kavli Institutes over the coming decade. This figure also includes the Foundation's continuing commitment to supporting project meetings and selected other activities.

<u>Salk Institute for Biological Studies</u>: The Salk Institute, under its Dynamic Brain Initiative, will dedicate over $28 million to work across traditional boundaries of neuroscience, producing a sophisticated understanding of the brain, from individual genes to neuronal circuits to behavior. To truly understand how the brain operates in both healthy and diseased states, scientists will map out the brain's neural networks and unravel how they interrelate. To stave off or

reverse diseases such as Alzheimer's and Parkinson's, scientists will explore the changes that occur in the brain as we age, laying the groundwork for prevention and treatment of age-related neurological diseases. (The White House, Office of the Press Secretary, "Fact Sheet: BRAIN Initiative," official website of the White House, last accessed October 4, 2013: http://www.whitehouse.gov/the-press-office/2013/04/02/fact-sheet-brain-initiative)

Now that you've read through that thrilling press release, let us give you a little insider information. Since we don't profess to be any kind of scientists, we sent another call for help to our science buddy Sharon Gilbert, who has a degree in biology and has spent decades reading published research into genetics (her specialty) and neuroscience (the subject of her unfinished doctoral work). She had this to say:

Tom and Terry, as you know, I've been following the brain-mapping hoopla since it was first announced last year, but I recently came across an article written by Giulio Prisco originally published at the website, Transhumanity—see: http://transhumanity.net/articles/entry/the-real-importance-of-brain-mapping-research, that sheds some much-needed light on the sudden race to map the brain. According to Prisco, the leap into brain mapping is actually based on a research paper published in the 2012 issue of *Neuron Magazine* called, "The Brain Activity Map Project and the Challenge of Functional Connectomics." Regarding this report, Prisco cites this from *Science Insider*:

In September 2011, George Church, the molecular geneticist who leads the Personal Genome Project, and Rafael Yuste, a neuroscientist at the Kavli Foundation and Columbia University, made waves at a meeting in England cosponsored by the Kavli, Allen, and Gatsby foundations when they proposed a massive, coordinated effort to develop technologies that can track the activity of functional connections in a living human brain, ultimately measuring "every spike from every neuron."

It's important that your readers know that two of the four listed authors of the BAM Project report (Brain Activity Map—BAM, which incidentally evokes oBAMa, does it not?) are Dr. George Church, a noted transhumanist and codirector of the Human Genome Project, and Rafael Yuste, a neuroscientist at the Kavli Foundation (one of the sponsors of the BRAIN project lauded and launched by the president). "Humans are nothing but our brains," Yuste said of the potential applications for technology produced in pursuit of a map of human brain activity. "Our whole culture, our personality, our minds, are a result of activity in the brain."

It certainly appears to me that both the European and the American brain projects intend to use the information to simulate a human mind in a silicon body. Prisco goes on in his article to reveal HIS beliefs in this chilling statement: *"Both BAM and HBP emphasize scientific results and medical* *applications, which of course are very important, but I hope that this race to the brain will produce real, game-changing break-throughs and take the first steps toward whole brain emulation, new mentality substrates, and **mind uploading**."* (Emphasis my own.) You can read the entire, revealing article at Turing-Church.com—http://turingchurch.com/2013/03/03/the-real-importance-of-brain-mapping-research/.

In short, the BRAIN project (following on the heels of the BAM project) and the Human Brain Project are presented as an Atlantic race for the neuro-prize, but in truth it is more likely a cooperative effort behind the scenes. Transhumanists MUST map the entire brain, all the intricate neurons and their millions of cellular connections, before attempting computer modeling and eventually "mind uploading." It's all about self-directed evolution. With Cold Spring Harbor Labs, HHMI, and DARPA all on the same bandwagon, you can bet that the Christians who oppose transhumanism are in for a bumpy ride.

Andrija Puharich and his cronies would have loved using today's supercomputers and brain-imaging software. Molecular Imaging Scans such as PET (positron emission tomography) and SPECT (single photon emission computed tomography) use magnetic resonance (MR) to create 3-D images of molecular reactions to specific probe molecules within your brain. Science now wants to do more than just "read our minds," it also wants to manipulate them. The voice-to-skull technology mentioned earlier is one crude way, but a report released in early August 2013 reveals a new method for "mind control":

Researchers from the Riken-MIT Center for Neural Circuit Genetics at the Massachusetts Institute of Technology took us closer to this science-fiction world of brain tweaking last week when they said they were able to create a false memory in a mouse.

The scientists reported in the journal *Science* that they caused mice to remember receiving an electrical shock in one location, when in reality they were zapped in a different place. The researchers weren't able to create entirely new thoughts, but they applied good or bad feelings to memories that already existed.

"It wasn't so much writing a memory from scratch, it was basically connecting two different types of memories. We took a neutral memory, and we artificially updated that to make it a negative memory," said Steve Ramirez, one of the MIT neuroscientists on the project.

It may sound insignificant and perhaps not a nice way to treat mice, but it is not a dramatic leap to imagine that one day this research could lead to computer-manipulation of the mind for things like the treatment of post-traumatic stress disorder, Ramirez said.

Technologists are already working on brain-computer interfaces, which will allow us to interact with our smartphones and computers simply by using our minds. And there are already gadgets that read our thoughts and allow us to do things like dodge virtual objects in a computer game or turn switches on and off with a thought.

But the scientists who are working on memory manipulation are the ones who seem to be pushing the boundaries of what we believe is possible. Sure, it sounds like movie fantasy right now, but don't laugh off the imagination of Hollywood screenwriters; sometimes the movies can be a great predictor of things to come.

In the movie, "Eternal Sunshine of the Spotless Mind," a character played by Jim Carrey uses a service that **erases memories to wipe his brain** of his former girlfriend, played by Kate Winslet.

But it seems the movie's screenwriter, Charlie Kaufman, was selling science short.

"The one thing that the movie 'Eternal Sunshine of the Spotless Mind' gets wrong, is that they are erasing an entire memory," said Ramirez of MIT. "**I think we can do better, while keeping the image of Kate Winslet, we can get rid of the sad part of that memory.**"

Hollywood and science-fiction writers, of course, have had fun with memory manipulation over the years.

In the film "Total Recall," which is based on a short story by Philip K. Dick, a character played by Arnold Schwarzenegger (and in a remake by Colin Farrell) receives a memory implant of a fake vacation to Mars. In "The Matrix," characters can download new skills like languages or fighting techniques to their mind, much like downloading a file to a computer.

Far-fetched? Perhaps, and we're not yet fighting our

robot overlords as the humans were in "The Matrix," but researchers really are exploring ways to upload new information to the brain.

In 2011, scientists working in collaboration with Boston University and ATR Computational Neuroscience Laboratories in Kyoto, Japan, published a paper on a process called Decoded Neurofeedback, or "DecNef," which sends signals to the brain through a functional magnetic resonance imaging machine, or FMRI, that can alter a person's brain activity pattern. In time, these scientists believe they could teach people how to play a musical instrument while they sleep, learn a new language or master a sport, all by "uploading" information to the brain.

Writing to the brain could allow us to interact with our computers, or other human beings, just by thinking about it.

In February, Dr. Miguel A. Nicolelis, a neuroscientist at Duke University, successfully connected the brains of two rats over the Internet, allowing them to communicate with their minds so when one rat pressed a lever, the other one did the same. The rats were in different locations, one at Duke University in North Carolina, and another in a laboratory in Natal, Brazil.

Nicolelis said he has recently performed other experiments in his lab where he has connected the brains of four mice in what he calls a "brain net" allowing them to share information over the Internet. In another experiment, he took two monkeys and gave them both half of a piece of information to successfully move a robotic arm, which required them to share the information through their brain. (all emphases added; "Scientists' Edge Closer to Manipulating Memory from a Computer in Brain," *The New York Times*, as quoted by *The Economic Times*, August 5, 2013, last accessed October 4, 2013: http://articles.economictimes.indiatimes.com/2013-08-05/news/41093205_1_memory-implant-brain-false-memory)

Did you catch the bit about "writing to our brain"? Who do these people think they are? God?

Well, yes.

Movies, television, and video games have been telling us for years that our every move is watched, our minds may not be our own, and even our memories are subject to question. Now, scientists are confirming that all we once believed to be fiction is, in reality: cold, hard FACT. The television program, *The X-Files*, is rife with end-times scenario programming and memes. The overarching story involves the planned invasion of the Earth by alien entities that reproduce via a virus contained within common oil. As the protagonist, Fox Mulder, seeks "the truth," he encounters a pilot from Area 51, who has been testing a spacecraft created from reverse-engineered alien tech. The episode is called "Jose Chung's From Outer Space" and is intended to keep the viewer guessing as to what is real and what is not. The plot follows a pair of teenagers who have been taken captive by aliens. Some scenes show the teens being hypnotized by their captors (aliens) while others reveal the hypnotist to be a military doctor. False memories twist the main characters' minds so that no one seems to know truth from trash. The scene with Mulder and the pilot takes place in a modest, 1950s-style diner:

JACK SCHAFFER: This is not happening! It's not happening! This is not happening. It's not happening. It's not happening.

(Later, they sit at a diner counter. Schaffer plays with his mashed potatoes with his fork. The cook is cleaning the counters near the front of the store. The pink neon sign shines brightly.)

Used to project the image of the Virgin Mary over the French trenches in World War One. The enemy's always willing to fire upon an invading force…but on a holy miracle?

MULDER: Or on visitors from outer space?

JACK SCHAFFER: Yeah, the enemy sees an American recon plane, they start shooting. They see a flying saucer from another galaxy…they hesitate.

(He puts down his fork.)

You know what happens to most people after seeing a UFO?

(He puts a cigarette in his mouth.)

MULDER: They experience "missing time."

JACK SCHAFFER: Any number of "soft option kills" will do…nerve gas…low frequency infrasound beams…

(He lights a match.)

With high-powered microwaves, you can not only cut enemy communications, you can cook internal organs.

(He lights the cigarette and shakes the match out.)

MULDER: But abductions?

JACK SCHAFFER: Don't know as much about them. I'm just the pilot. You ever flown a flying saucer?

(Mulder shrugs slightly.)

Afterwards, sex seems trite.

(He takes another drag.)

MULDER: But what do you do with the abductees?

JACK SCHAFFER: Take them back to the base. Let the doctors work on them. Nothing physical, they just mess with their minds.

MULDER: Hypnosis.

(Schaffer nods.)

JACK SCHAFFER: At the base, I've seen people go into an ordinary room with an ordinary bunch of doctors…and come out absolutely positive they were probed by aliens.

MULDER: But if abductions are just a covert intelligence operation and UFOs are merely secret military airships, piloted by aliens such as yourself…then what were you abducted by?

JACK SCHAFFER: Don't you get it? I'm absolutely positive me, my copilot, and those two kids were abducted but I can't be absolutely sure it happened. I can't be sure of anything anymore!

MULDER: What do you mean?

JACK SCHAFFER: I'm not sure we're even having this conversation. I don't know if these mashed potatoes are really here. I don't know if you even exist.

MULDER: I can only assure you that I do.

JACK SCHAFFER: Well…thanks, buddy. Unfortunately…I can't give you the same assurance about me.

(The door slams. Mulder and Jack look over to see a number of soldiers being led by the Air Force Man. Jack takes a drag of his cigarette and prepares to leave.)

JACK SCHAFFER: Well, looks like I'm a dead man. (Script for "Jose Chung's From Outer Space," *X-Files*, as quoted by *Inside the X*, last accessed October 4, 2013: http://www.insidethex.co.uk/transcrp/scrp320.htm)

Dead men tell no tales—or so we're told. But an uploaded memory is there for all to enjoy. In many ways, our memories define us. Alzheimer's patients lose track of "self" because they've lost access to important, self-describing memories. Transhumanists do not think of you and me as unique, God-created, God-designed individuals—rather, they see us as machines made of flesh. Our memories are just data files that can be corrupted, manipulated, altered, or *erased*.

To "Enhance" Mankind Physically, Neurologically, and Spiritually

Are we a spirit, a soul, a conscious to be transferred at will? In February 2011, the "Darwin Day" Forum—a gathering of scientists and transhumanists in honor Charles Darwin—met in New York City to discuss the nature of human consciousness. One presentation featured New York University (NYU) philosopher Ned Block and neuroscientist Jacqueline Gottlieb from Columbia University. The event was held at the Tishman Auditorium at NYU. Block believes that our consciousness involves visual field phenomena. It's a neuro-philosophical version of Shrodinger's dead/alive cat. We are defined by what we perceive, and our world is therefore represented by what we "choose" to see. Block posits that reality *is* what we choose to see—raising the question of whether or not something exists if we do not perceive it. By using eye-tracking software, Block observed the perception phenomenon for various subjects. Over and over, many subjects overlooked the same areas—effectively making that area of the field of vision "invisible."

Working memory for Block is defined by those things that we have perceived and can recall. His experiments with macaque monkeys led him to posit that human brains have only so many "slots" in them for memory. It's comparable to the memory limits for cache or RAM in computers. Again, here is a scientist who not only compares humans to monkeys, but also appears to see our brains as little more than living computers.

Are we sentient computers? The hit television series *Battlestar Galactica* follows a band of humans who are ruthlessly pursued by sentient computers called Cylons. In the ending, it's revealed that some of the humans are actually Cylons, and that these "final five" (a God-like hand of five fingers who create) are the ones who, in best Promethean style, gave a formerly defeated band of Cylons the secret to appearing human *and* the secret of regeneration (eternal life). The final question one must ask then is: Are the ragtag band of humans really human—or are all humans merely sentient machines?

Transhumanists predict that humanity will eventually blend with and give rise to sentient machines. In his August 5, 2013, article at IEET (Institute for Ethics and Emerging Technology), George Deane explores the idea of "sentient computers" by answering an earlier post by John R. Searle:

John Searle delivered a powerful blow to computationalism by debunking the notion that syntax is sufficient for semantics with the Chinese Room Argument, but he later went further to argue that the theory is not only false but also incoherent.

To understand Searle's point it is first worth making a distinction between features of the world that are independent of the observer, i.e., intrinsic and not pliable to interpretation, and those which are relative to the observer, or extrinsic. Typically the natural sciences are concerned with the former. It is worth noting observer relative features of the world can also be objective; many objects such as wallets, books and clothing are not defined in terms of physics but in terms of function. Armed with this distinction we can ask, is computation observer relative or observer dependant [*sic*]?...

Searle puts this point best of all: "For any program there is some sufficiently complex object such that there is some description of the object under which it is implementing the program. Thus for example the wall behind my back is right now implementing the Wordstar program, because there is some pattern of molecule movements which is isomorphic with the formal structure of Wordstar. But if the wall is implementing Wordstar then if it is a big enough wall it is implementing any program, including any program implemented in the brain."

Unless we accept that a large enough wall would be instantiating every possible state of consciousness simultaneously then it appears there is more to consciousness than computation. If computation can [be] attributed to anything then it would be trivial to call the brain a computer. Searle proposes we look beyond computation to what the brain actually is: a physical system. This is a damaging blow to accounts of mind uploading and Strong AI that ignore the substrate of implementation. Those hoping to achieve digital immortality through whole brain emulation might not want to be too hasty to dispose of their biological bodies after all. Sharper criteria for consciousness need to be defined. (George Deane, "Can a Computer Have Consciousness?" *IEET*, July 31, 2013: http://ieet.org/index. php/IEET/more/deane20130805)

It's nice to know that not all transhumanists believe we're within minutes of uploading. Deane, who is studying in London, describes himself as "especially interested in Neuroethics and the implications of technologies for **cognitive enhancement**" (ibid.; emphasis added).

Let's explore that concept for a moment. According to the Center for Neuroscience and Society's website (http://neuroethics.upenn.edu/index.php/penn-neuroethics-briefing/cognitive-enhancement):

Two main cognitive systems have been targeted for **pharma-cological enhancement**: attention and memory. Stimulant drugs such as methyphenidate (Ritalin) and amphetamine (Adderall) improve the attention of people with attention deficit hyperactivity disorder (ADHD) and can also enhance attention in healthy people. (emphasis added)

Pharmacological enhancement? Is it possible that the mountain of psychotropic drugs being tested in prisons, schools, and nursing homes across the world constitute a massive study group investigating pharmacologically enhanced cognition? This echoes the military assignment given to Captain Andrija Puharich to find a chemical means to not only alter the mind but open it so that the psyche, the consciousness, is released. Puharich and The Nine and the *Star Trek* world of make-believe have informed our reality. You and I now live in a world effectively designed by Gene Roddenberry. We carry sensors, cameras, recording devices, and instantaneous communicators—all with the help of a spidery connection called the Internet or World Wide Web.

The television program *X-Files* had an "end-date" of December 21, 2012, when the aliens would land and take over the world. Only one place was considered safe—and it is there that the Cigarette Smoking Man (Mulder's true biological father) ran for safety. This area lies in the four-corners region of the American Southwest, where the Hopi now live—the land the Hopis claim was once inhabited by the Anasazi.

The Hopi eschatology goes like this:

The following extraordinary Hopi prophecy was first published in a mimeographed manuscript that circulated among several Methodist and Presbyterian churches in 1959. Some of the prophecies were published in 1963 by Frank Waters in *The Book of the Hopi*. The account begins by describing how, while driving along a desert highway one hot day in the summer of 1958, a minister named David Young stopped to offer a ride to an Indian elder, who accepted with a nod. After riding in silence for several minutes, the Indian said:

> "I am White Feather, a Hopi of the ancient Bear Clan. In my long life I have traveled through this land, seeking out my brothers, and learning from them many things full of wisdom. I have followed the sacred paths of my people, who inhabit the forests and many lakes in the east, the land of ice and long nights in the north, and the places of holy altars of stone built many years ago by my brothers' fathers in the south. From all these I have heard the stories of the past, and the prophecies of the future. Today, many of the prophecies have turned to stories, and few are left—the past grows longer, and the future grows shorter.
>
> "And now White Feather is dying. His sons have all joined his ancestors, and soon he too shall be with them. But there is no one left, no one to recite and pass on the ancient wisdom. My people have tired of

the old ways—the great ceremonies that tell of our origins, of our emergence into the Fourth World, are almost all abandoned, forgotten, yet even this has been foretold. The time grows short.

"My people await Pahana, the **lost White Brother** [a being, supposedly from the stars], as do all our brothers in the land. He will not be like the white men we know now, who are cruel and greedy. We were told of their coming long ago. But still we await Pahana.

"He will bring with him the symbols, and the missing piece of that sacred tablet now kept by the elders, given to him when he left, that shall identify him as our True White Brother.

"The Fourth World shall end soon, and the Fifth World will begin. This the elders everywhere know. The Signs over many years have been fulfilled, and so few are left.

"This is the First Sign: We are told of the coming of the white-skinned men, like Pahana, but not living like Pahana men who took the land that was not theirs. And men who struck their enemies with thunder.

"This is the Second Sign: Our lands will see the coming of spinning wheels filled with voices. In his youth, my father saw this prophecy come true with his eyes—the white men bringing their families in wagons across the prairies.

"This is the Third Sign: A strange beast like a buffalo but with great long horns, will overrun the land in large numbers. These White Feather saw with his eyes—the coming of the white men's cattle.

"This is the Fourth Sign: The land will be crossed by snakes of iron [the railroads].

"This is the Fifth Sign: The land shall be crisscrossed by a giant spider's web [the Internet].

"This is the Sixth sign: The land shall be crisscrossed with rivers of stone that make pictures in the sun.

"This is the Seventh Sign: You will hear of the sea turning black, and many living things dying because of it.

"This is the Eighth Sign: You will see many youth, who wear their hair long like my people, come and join the tribal nations, to learn their ways and wisdom.

"And this is the Ninth and Last Sign: You will hear of a dwelling-place in the heavens, above the earth, that shall fall with a great crash. It will appear as a blue star. Very soon after this, the ceremonies of my people will cease [this may refer to the Blue Kachina].

"These are the Signs that great destruction is coming. The **world shall rock to and fro** [reminiscent of the Bible's prophecy in Isaiah 24:20 that the Earth shall reel like a drunkard]. The white man will battle against other people in other lands—with those who possessed the first light of wisdom. There will be many columns of smoke and fire such as White Feather has seen the white man make in the deserts not far from here. Only those which come will cause disease and a great dying. Many of my people, understanding the prophecies, shall be safe. Those who stay and live in the places of my people also shall be safe. Then there will be much to rebuild. And soon—very soon afterward—Pahana will return. He shall bring with him the dawn of the Fifth World. He shall plant the seeds of his wisdom in their hearts. Even now the seeds are being planted. These shall smooth the way to the Emergence into the Fifth World." ("The Hopi Prophecy," viewable online, last accessed October 4, 2013: http://www. welcomehome.org/rainbow/prophecy/hopi1.html)

The Pahana mentioned above refers to the Hopi messiah, a man from the stars who will bring with him an artifact to prove his position. This man could easily be the coming Antichrist. Some call him Mahdi; Buddhists look for Maitreya; Hindus for Kalki; Zoroastrians for Saoshyant; Nostradamus may have called him Mabus. All these refer to the Man of Sin—the coming world ruler who will call himself Christ, Messiah, Maitreya, and perhaps Pahana. Reality as we know it may become nothing more than a Matrix of perception. Our minds may hear instructions or enticements in our own language, while our eyes see only what we wish to see. We have heaped unto ourselves false teaching because of our itching ears, and this is what God has permitted: A strong delusion.

Metal will meld with flesh and mind with falsehoods as those who do not choose the true Christ, Jesus of Nazareth, are transformed into the transhumanist zealots' twisted image of godhood. Human Artificial Chromosomes, HACs, may be used to mark us—as may tattoos embedded with genetic information. We shall all be changed—some to immortality with Christ, some to hideous clay and iron constructs that mock God's beautiful design by attempts to "enhance" His code. Cloning, human/animal hybridization, mind control, pharmaceuticals, and the false notion of self-directed evolution are all part of an end game that the enemy hopes will allow him to defeat God. The final armies will gather in Jerusalem—not to fight each other, but to look up! The true Christ will return at the end of this age, and He will do battle with Earth's demon-led enhanced human army.

We know the ending. We know the enemy is already defeated—he just won't admit it…*yet!*

chapter eight

Trigger Event—When 666 Becomes Mandatory Overnight

In previous chapters, we have cited the book of Revelation (13:16): "And he *causeth* all, both small and great, rich and poor, free and bond, to receive a mark in their right hand, or in their foreheads: And that no man might buy or sell, save he that had the mark, or the name of the beast, or the number of his name" (emphasis added).

The particular phrasing in this verse is very important because the Greek verb for "causeth" (*poieō*) implies that something is set in motion by the instigator of an event (in this case, the Antichrist), which then triggers others to have to respond to it.

One wonders just what the Antichrist will initiate that will result in the majority of the global population making a decision to accept his Mark. It would have to be something extraordinary to put so much pressure on freedom-loving peoples around the world—especially in Judeo-Christian cultures—to cause them to lay aside personal freedoms and eternal salvation in exchange for an Orwellian society where One World Government oversees the smallest details of their lives and in which human liberty is abandoned.

Given the cherished idea of "free moral agency," this will be no easy task, as our friend Noah Hutchings of Southwest Radio Ministries noted in an article he wrote and recently sent to these authors, a portion of which stated:

> God created man in His own image, and He instilled within the spirit of man the will to be a free moral individual. Man has the election to obey the Creator or disobey Him and follow the dictates of his conscience. [Man also] has the inherent right to select the type of government that is to regulate the order of society. Countless wars have been fought and millions have died to protect this freedom. Liberty has been subjected to the will of tyrants and dictators, yet it has surfaced down through the ages and endured as the hope of mankind. The Pilgrims came to America to find a haven for personal liberty. The Revolutionary War was fought to preserve this freedom. The Civil War was fought to bring freedom to those who were enslaved and without liberty. Thomas Jefferson wrote in the Declaration of Independence: "All men are created equal, and endowed by their Creator with life, liberty, and the pursuit of happiness." But a time is coming, and we believe it is very near, when individual freedom will be completely extinguished: "…and power was given him

over all kindreds, and tongues, and nations. And all that dwell upon the earth shall worship him, whose names are not written in the book of life of the lamb slain from the foundation of the world" (Revelation 13:7–8).

Will the eradication of the spark of liberty within the spirit of man be accomplished by a tyrannical system of government like communism or fascism? Will it be because men surrender individual freedom in order to survive the threat of international terrorism, nuclear war, or some other imminent danger that threatens the very existence of mankind? Or will it come about because…the masses [have] become so conditioned to state supervision and control…that the vast majority will accept the mark and number of the Beast when it arrives[?] The links in the chain of bondage which the Antichrist will use when all the world becomes a prison are being forged. A letter by Dr. Michael W. Fox, scientific director of the Humane Society of the United States, written to the Society for the Protection of the Individual Rights and Liberties of Tucson, Arizona, states in part:

> You may be interested to know that one military PX, I believe in Ft. Worth, is considering inserting electronic identification in the back of the hands of people using the PX to permit them ready entry without any other screening. It seems as though this age of biotechnology is being used for greater control and manipulation, and the pyramid of power is being strengthened in the process.

The course of events in technology, science, economics, crime and social behavior, nuclear war, and international terrorism is bringing mankind to [a trigger event] to take the mark and number of Antichrist or be killed. How near is the present generation to facing this decision? The developing bondage of the human body, mind, and spirit is evidence that it *is* near.

As usual, Dr. Hutching's insights are acutely tuned and do not fail to recognize the need for a global-scale incident to set in motion the Man of Sin's "cause," which then somehow will overwhelm the average person's ability to resist receiving the Mark of the Beast.

What could this far-reaching scenario conceivably be? While the prospects are numerous, for the sake of space let us describe just three possibilities: 1) an EMP (electromagnetic pulse) or cyber attack; 2) a global economic meltdown; and 3) a natural or manufactured global pandemic, the development of which these authors believe could most directly connect to a beastly Mark.

Scenario #1: Electromagnetic and/or Cyber Warfare Event

Most government experts in fields of risk mitigation are focused currently on likely terrorist or enemy-state-sponsored scenarios that could take advantage of aging electrical and water systems infrastructure. For example, a rogue nation could load "defensive" missiles with nuclear warheads and launch them as multiple electromagnetic pulse weapons (HEMPs or EMPs) above the United

States and/or other countries from offshore freighters. In a literal "flash," this could bring down national electrical grids as well as telecommunications; critical energy resources such as oil and natural gas pipelines; water delivery systems; banking and financial institutions, including consumer transactions; and emergency services. In military terminology, if a nuclear warhead is detonated hundreds of kilometers above the Earth's surface in this way, it is known as "a high-altitude electromagnetic pulse (HEMP) device. Typically the HEMP device produces the EMP as its primary damage mechanism. The nuclear device does this by producing gamma rays, which in turn are converted into EMP in the mid-stratosphere over a wide area within line of sight to the detonation. NEMP is the abrupt pulse of electromagnetic radiation resulting from a nuclear explosion. The resulting rapidly changing electric fields and magnetic fields may couple with electrical/electronic systems to produce damaging current and voltage surges" ("Electromagnetic Pulse," *Wikipedia: The Free Encyclopedia*, last modified October 4, 2013, last accessed October 4, 2013: http://en.wikipedia.org/wiki/Electromagnetic_pulse).

On this order, warnings have been issued recently by the US Defense Department, Homeland Security, and even President Obama involving the risk associated with enemy nations that are already "testing" US electric infrastructure vulnerability through cyber attacks aimed at determining how to sabotage the power grid, financial institutions, and even air traffic control systems.

But that may not be the worst of it, according to a February 26, 2013, investigative report by F. Michael Maloof for *World Net Daily*. As a former senior security policy analyst in the office of the secretary of defense and a current staff writer for *WND* and *G2Bulletin*, Maloof points out in *Sledgehammer of Cyber Warfare? EMP Attack* that:

> Those same adversaries—China, Russia, Iran and North Korea—also incorporate in their military doctrine the use of a nuclear electromagnetic pulse, or EMP, attack as "part of a strategic operation that would basically 'throw the kitchen sink' at the United States," according to Cynthia E. Ayers, who once was with the National Security Agency and currently is with the US Army War College.
>
> These countries, she said, will "hit us with everything—computer viruses, sabotage of critical communications nodes, kinetic strikes on key information systems and a nuclear EMP attack."
>
> "The last, an EMP, is their best chance to collapse our national power grid and take us down, perhaps permanently," she said.

In short, a coordinated assault on America and/or allied nations as described above could instantly result in the collapse of Western society as it has been known. Directly thereafter, anarchy would fill the streets, martial law would be imposed, and a national cry would fill the air for salvation from chaos.

But is this a perfect recipe through which the Man of Sin could step in with wondrous answers to our overnight problems by offering some unknown method for restoring social order? How could he "mark" those who would be allowed to function in his kingdom following a catastrophic cyber or EMP attack resulting in widespread damage to essential structures? Wouldn't electrical and computerized systems be needed online for implantable or high-tech "marks" to function as we have described them? Surely global superpowers like the United States have top-secret responses already devised for just such an event, but perhaps something beyond human comprehension is planned by the Antichrist. Intriguingly, when the Bible describes how the final world leader will deceive the world with "lying signs and wonders" (2 Thessalonians 2:9) and appear at a time when there are "fearful sights and great signs from heaven" (Luke 21:11), one cannot help but note the specific and interesting wording that implies how the Man of Sin will be directly associated with *electricity and people's need of shelter.*

First, the "heaven" mentioned in Luke 21:11 is not the throne room of God, but rather *ouranos,* or the vaulted expanse of the sky, where the clouds and the tempests gather and where electrical energy or "lightning" is produced.

Note that in Nehemiah 9:6, the prophet spoke of more than one heaven: He saw the heavens and the "heaven of heavens." Paul also referred to different "heavens" in 2 Corinthians 12:2, saying, "I knew a man in Christ above fourteen years ago, (whether in the body, I cannot tell; or whether out of the body, I cannot tell: God knoweth;) such an one [was] caught up to the third heaven." Some scholars believe when Paul referred to this third heaven, he was echoing his formal education as a Pharisee concerning three heavens that included a domain of air (the *kosmos*) or height, controlled by Beelzeboul (Satan), the "lord of the height" and god of lightning (electricity). In pharisaical thought, the first heaven was simply the place where the birds fly— anything removed from and not attached to the surface of the Earth. On the other end of the spectrum and of a different substance was the third heaven—the dwelling place of God. Between this third heaven, "where dwells the throne room of God," and the first heaven, where the birds fly, was a war zone called the "second heaven." This was *ouranos,* or as it was also known, the *kosmos*—the Hebrew equivalent of the Persian *Ahriman-abad*—the place where Satan abides as the prince of the power of the "air" (Greek *aer:* the lower air, circumambient, location of natural electrical energy), a sort of gasket heaven, the domain of Satan encompassing the surface of the Earth.

In Persian theology, the spirit that opposed the prophet Daniel (see Daniel 10) and his angel would have been identified as this same spirit, but called *Ahriman* in their culture, a god capable of electricity and lightning whose legend closely parallels the biblical fall of Lucifer. According to Persian religion, Ahriman was the death dealer—the powerful and self-existing evil spirit from whom war and all other evils had their origin. He was the chief of the cacodemons, or fallen angels, expelled from heaven for their sins. After being kicked out of heaven, the cacodemons endeavored to settle down in various parts of the Earth, but were always rejected, and out of revenge found pleasure in tormenting the inhabitants of the Earth. Ahriman and his followers finally took up their abode in the space between heaven and Earth and there established their domain: *Ahriman-abad*—"the abode of Ahriman." From this location, the cacodemons could intrude into and attempt to corrupt the governments of men.

In the Bible, both the False Prophet and the Antichrist are

described as being aligned with the "power" of this celestial realm, from which they are able to call down "fire," presumably lightning or electricity (though the reference could actually be of literal fire).

The prophet Daniel also tells us the Antichrist's belief system will honor a "strange, alien god" who again appears related to electricity and lightning. In Daniel 11:38–39, we read:

> But in his estate shall he honour: and a god whom his fathers knew not shall he honour with gold, and silver, and with precious stones, and pleasant things. Thus shall he do in the most strong holds with a , whom he shall acknowledge and increase with glory: and he shall cause them to rule over many, and shall divide the land for gain. (emphasis added)

Several parts of Daniel's prophecy stand out as very unusual. First, the phrase, "God of forces,"—or, alternately, "god of fortresses"—has been connected to Baal-Shamem, an ancient deity who was worshipped throughout the Middle East, especially in Canaan/Phoenicia, Syria, and later by the Manichaean Gnostics, who revered him as the greatest angel of electrical energy, or natural, electrical, and high-voltage discharges (lightning). Could this imply that the Antichrist will come on the world scene overnight like some supernatural version of Nikola Tesla, with a wonder-creating wireless electricity transmission system capable of repowering the world?

(Note: The nineteenth-century inventor, Nikola Tesla, was an electrical engineer known around the world for his patented devices and contributions to the knowledge of alternating current

[AC] electricity supply systems. Before he died in 1943, Tesla had intended a proof-of-concept demonstration of an intercontinental system, which would provide transatlantic wireless power transmissions for electricity, telephones, and broadcasting, but the project was defunded midstream by pressure brought on by—some believe—skullduggery involving electricity contracts competitor General Electric, Thomas Edison's company.)

In addition to a wireless electricity, note that the Hebrew word translated "forces" in this verse is *ma`owz*, which refers to a deity who can provide human protection in the form of strong housing, places of safety, protection, and refuge—adding to the question of whether Antichrist could appear as a false savior of humanity immediately following a wide-scale cyber attack or EMP-type event that brings down national power grids, leaving people desperate for common necessities such as electricity and housing.

Finally, connected to the appearance of the Beast and Antichrist, the book of Revelation seems to preface that moment with a military attack resulting in the destruction, or "death," and then seemingly miraculous rapid recovery of what many expositors believe is both a man and the global supernation he represents. According to this premise, the healing of that man and/or nation's "deadly wound" causes all the world to worship "the dragon which gave *power* unto the beast" and to proclaim: "Who is like unto the beast? who is able to make *war* with him?" (emphasis added; Revelation 13:4). This is then followed by a vision of "power" and "fire" from the heavens connected to the implementation of the Mark of the Beast (verses 13–18):

And he doeth great wonders, so that he maketh fire [electricity?] come down from heaven on the earth in the sight of men, And deceiveth them that dwell on the earth by the means of those miracles which he had power to do in the sight of the beast; saying to them that dwell on the earth, that they should make an image to the beast, which had the wound by a sword, and did live. And he had power to give life unto the image of the beast, that the image of the beast should both speak, and cause that as many as would not worship the image of the beast should be killed. And he causeth all, both small and great, rich and poor, free and bond, to receive a mark in their right hand, or in their foreheads: And that no man might buy or sell, save he that had the mark, or the name of the beast, or the number of his name. Here is wisdom. Let him that hath understanding count the number of the beast: for it is the number of a man; and his number is Six hundred threescore and six.

Scenario #2: The Collapse of the Global Economy

Recently, the United States' National Intelligence Council (NIC) and the European Union's Institute for Security Studies (EUISS) joined forces to produce an assessment of the long-term prospects for global governance frameworks. The report—*Global Governance 2025: At a Critical Juncture*—assessed leading intercontinental perils that could endanger the collective administration of shared problems at the international level. From the beginning of the report under the subsection "Scenario I: Barely Keeping Afloat," the writers acknowledge how crises including current financial institutions are being served ad hoc, temporary frameworks devised to avert the most threatening aspects (such as the United States printing money for which it has no gold reserves) and synthetic economic tricks being used to temporarily sustain what is ultimately an unsustainable financial system.

Conservative analysts have been predicting a devastating crash of the stock market for quite some time, all the while holding their breath, hoping it won't happen. Yet history and experience, coupled with extraordinary facts available today, convince them that the world's major economies are being artificially sustained and that it is only a matter of time before this house of cards collapses. Even more conspiratorially, some suggest something sinister is actually being *planned,* as in a global stock market crash for the near future unlike anything the world has experienced before. Such a crash would permit the Illuminists and cohorts to close thousands of banks in a matter of days, seize most personal assets, confiscate gold and silver, and eliminate cash, all under federally sanctioned "declared emergencies" activated by presidential executive order. After the financial institutions of the world crumble, a worldwide monetary system would be restructured into one that provides more efficient methods of total enslavement, setting the stage for the official establishment of the Antichrist's New World Economic Order.

This manufactured crash might begin in a country like Japan and then work its way around the globe, toppling the economies of nations like a row of dominoes, virtually simultaneously.

On the heels of this event, a new form of digital currency could be announced that is international in scope and proclaimed as more reliable than the old monetary system. It would replace modern credit and debit cards as well as paper checks, ultimately paving the way for a super biometric ID card, smart tattoo, or biochip implant wherein every financial transaction in one's life could be stored, catalogued, analyzed, and accessed for future reference by New World Order bureaucrats. For the majority of people who have been using electronic banking for years (via direct payroll deposits, direct deposit of Social Security checks, ATMs, credit and debit cards, electronic automatic payments of bills, etc.), accepting the new system will be a snap.

What's more, as detailed in earlier chapters, polls around the world show overwhelmingly today that a majority approves and appreciates the convenience of emerging biometric and "smart" banking technologies and are open to near-future realities wherein their flesh will be merged with apparati for buying and selling (and surveillance), either via an implantable chip or some other cool, new, cyborg control system.

What's that?

Cyborgs?

New people for a new system?

Are you serious?

Apparently.

When the *Global Governance 2025: At a Critical Juncture* document cited above combined the risks of a global financial collapse triggered (in one scenario) by "biological weapons," it included in that assessment how biotech could also eventually lead to *a new form of man*—a previously unknown human with unique physical, emotional, and cognitive abilities that emerge as a result of the very science and technology the Antichrist may use to enslave humanity.

Note that on page 35 of their report, these top-shelf United States and European intelligence leaders transition from the threat of a biological weapon *to the creation of a potentially dangerous new form of man*:

> No forum currently exists for dealing comprehensively across the scientific community, industry, and governments on measures needed to diminish the risks posed by the bio-technology revolution. The development of new agents and the expansion of access to those with hostile intentions increase the bioterrorism threat.… In addition, biotech-nology—which the OECD thinks will potentially boost the GDPs of its members—can drive *new forms of human* behavior and association, creating profound cross-cultural ethical questions that will be increasingly politically conten-tious. Few experts believe that current governance instru-ments are adequate for those challenges. For example, *direct modification of DNA* at fertilization is widely researched with a goal of removing defective genes; however, discus-sions of future capabilities *open the possibility for designing humans with unique physical, emotional, or cognitive abilities.* (emphasis added)

Bio-enhanced humans with unique "cognitive" abilities have been in the design budgets and on the drawing board of military strategists and social engineers who, for some time, have imagined how man's growing marriage with—and dependence on—machine intelligence will, in the not-too-distant future, accompany almost everything we do, including buying and selling. The era has already started, though it is in its embryonic stage, and it has been given a name. It is being called the "Hybrid Age." And yes, this means exactly what it sounds like. What we are already doing with genetically modified crops, transgenic animals, and human-animal chimeras at the embryonic scale, we intend to do to the rest of humanity in general—to hybridize man via genetic alterations, nanotechnology, synthetic biology, and human-tech integration with artificial intelligence and brain-machine interfaces.

Parag Khanna and Ayesha Khann explain in their August 19, 2011, article "Foreign Policy: A Predictable Future for Technology":

> As we try to understand an incipient future in which technology has insinuated itself into every sphere and nook of human activity—from the manipulation and replication of DNA to space exploration—and in which humans continuously seek ways to speed up their biological evolution to match the breakneck pace of technological evolution, the only way to do that is to incrementally integrate with technology, launching an era of change and innovation that we call the Hybrid Age. If the first wave was agrarian and tribal, the second industrial and national, and the third informational and transnational, then the Hybrid Age is…the "Fourth Wave." In this new era, human evolution [will] become human-technology co-evolution:

> We're becoming part of the machine, and it is becoming part of us. (Parag Khanna and Ayesha Khann, "Foreign Policy: A Predictable Future for Technology," *NPR*, August 19, 2011: http://www.npr.org/2011/08/19/139779301/foreign-policy-a-predictable-future-for-technology)

If the description above sounds like an incredible dream (or nightmare), consider how quickly such technology is spreading into the broader culture (and now with extracranial applications, it will no longer require the Beast chip surgically implanted into one's brain). Currently, tests are being conducted that allow people to interact with computers, smart phones, and tablets simply by using their minds. Games systems have been on the market since 2010 that let players control some of the functions with "thought" alone by wearing a rubber cap that reads and translates their brain's electrical impulses. More advanced systems under development at the University of Washington allow one person to send a brain signal over the Internet to a second gamer across campus in a different building who is wearing such a "cap." The second gamer "involuntary" clicks a tab with his index finger as he receives signals from the first person, who is playing the game by using the second person's mind. Over the next few years, people everywhere will be turning on their lights at home, sending emails, and yes, soon thereafter, transferring monetary funds by thought-controlled, brain-machine interfacing without ever pulling their wallets or checkbooks from their pockets or purses. Welcome to the Borg.

In the April 28, 2013, *New York Times* article, "Disruptions: Brain Computer Interfaces Inch Closer to Mainstream," Nick

Bilton (http://bits.blogs.nytimes.com/2013/04/28/disruptions-no-words-no-gestures-just-your-brain-as-a-control-pad/) elaborates:

> But that chip inside the head could soon vanish as scientists say we are poised to gain a much greater understanding of the brain, and, in turn, technologies that empower brain computer interfaces. An initiative by the Obama administration this year called the Brain Activity Map project, a decade-long research project, aims to build a comprehensive map of the brain.
>
> Miyoung Chun, a molecular biologist and vice president for science programs at the Kavli Foundation, is working on the project and although she said it would take a decade to completely map the brain, companies would be able to build new kinds of brain computer interface products within two years.
>
> "The Brain Activity Map will give hardware companies a lot of new tools that will change how we use smartphones and tablets," Dr. Chun said. "It will revolutionize everything from robotic implants and neural prosthetics, to remote controls, which could be history in the foreseeable future when you can change your television channel by thinking about it."

Thus the technology for the Beast's end-times system is practically here. All the Antichrist will need is the trigger event, which could come in the form of a global financial meltdown that brings the world to its knees and introduces a modern and more secure method of buying and selling via high-tech "marks." Highly educated economists around the world say we are on the precipice of that cascading event now.

Scenario #3: Natural or Manufactured Global Pandemic

As noted in an earlier chapter on RFID technology, scannable implants and tattoo transmitters are becoming more sophisticated, adding "prophetic" components—such as merging human biological matter with transistors to create living, implantable machines. Related science also envisions "smart" chimeric vaccines that literally rewrite DNA. The authors of this book believe the possibility that the Mark of the Beast could arrive through a version of one of these technologies is plausible, if not altogether likely.

But again, what "trigger" could set in motion the need for a universal vaccine in the form of a biochip?

This is an important question, because a while back, one of these author's wives (Nita Horn) brought up a point we had never considered. She asked if the biblical Mark of the Beast might be a conspiracy employing specific implantable technology only now available. Her theory was gripping: An occult elite operating behind the US government devises a virus that is a crossover between human and animal disease—let's say, an entirely new and highly contagious influenza mutation—and intentionally releases it into the public. A pandemic ensues, and the period between when a person contracts the virus and death is something like ten days. With tens of thousands dead in a few weeks and the rate of death increasing hourly around the globe, a universal cry for a cure goes out. Seemingly miraculously, the government then steps forward with a vaccine. The only catch, the government explains, is that, given the nature of the animal-human flu, the "cure" uses animal DNA and nanobots to rewrite one's genetics so that the person is no longer entirely human. The point made was that those who receive this antidote would become part "beast," and perhaps, thus, the title, "Mark of the Beast." No longer "entirely human" would also mean—according to this outline—that the individual could no longer be "saved" or go to heaven, explaining why the book of Revelation says that "whosoever receiveth the mark" is damned forever (while also explaining why the ancient *nephilim*, whose DNA was part human and part angel and/or animal, could not be redeemed). For believers, they would know this is the Mark of the Beast, as we believe the Spirit of God would confirm it in their conscious and Spirit the true facts. If they take this "cure," they will know at the moment they have willfully submitted to damnation regardless how it is justified.

If one imagines the global chaos of such a pandemic, the concept of how the Antichrist "causes all," both "small and great," to receive this Mark becomes clearer. When looking into the eyes of dying children, parents, or a spouse, it would be incredibly difficult to allow oneself to die or to encourage others to do the same when a "cure" was readily available. Lastly, this scenario would mean that nobody would be allowed to "buy or sell" in the marketplace without the Mark-cure due to the need to quarantine all but the inoculated, thus fulfilling all aspects of the Mark of the Beast prophecy.

To find out if the science behind this abstract was as reasonable as it appeared on the surface, we contacted Sharon Gilbert, whose graduate work included molecular biology (as mentioned prior). This was her troubling response:

> What is human? Until recently, most of us would readily respond that *we* are humans. You and I, we might argue, are *Homo sapiens*: erect, bipedal hominids with twenty-three pairs of matched chromosomes and nifty little thumbs capable of apposition to the palm that enable us to grasp the fine tools that our highly developed, bi-lobed brains devise.
>
> Humans, we might argue, sit as rulers of the Earth, gazing down from the pinnacle of a pyramid consisting of all plant and animal species. We would remind the listener that natural selection and evolution have developed mankind into superior thinkers and doers, thereby granting us royal privilege, if not infinite responsibility. The Bible would take this definition much farther, of course, adding that mankind is the only part of God's creation formed by His hands, rather than spoken into existence, and that you and I bear God's unique signature as having been created "in His image" (Genesis 1:27).
>
> Many members of the "illuminated brotherhood of science" would likely demur to the previous statement. These have, in point of fact, redefined *human*. Like [Mary] Shelley's *Modern Prometheus*, Victor Frankenstein, today's molecular magicians play "god" not by stitching together rotting corpses, but by reforming the very essence of our beings: our DNA.

So-called "postmodern man" began as a literary reference, but has evolved into an iconic metaphor representing a collective image of perfected humanity beyond the confines of genetic constraints. Transhumanism, also known as the H+ movement (see www.HPlusMagazine.com, for example) envisions a higher life-form yet, surpassing *Homo sapiens* in favor of *Homo sapiens 2.0*, a bioengineered construct that fuses man's original genome with animal and/or synthetic DNA.

While such claims ring of science fiction, they are indeed science fact. For decades, laboratories have created chimeric combinations of animal, plant, and even human DNA under the guise of medical research. The stated goal is to better man's lot by curing disease, but this benign mask hides an inner, sardonic grin that follows an ancient blueprint to blend God's perfect creature with the seed of fallen angels: "You shall be as gods."

You two speak to the heart of the matter when you warn of a day when true humans may receive transhuman instructions via an implant or injection. A seemingly innocuous vaccine or identification "chip" can initiate intracellular changes, not only in somatic or "body" cells, but also in germ-line cells such as ova and sperm. The former alters the recipient only; the latter alters the recipient's doomed descendents as well.

In my second novel, *The Armageddon Strain*, I present a device called the "BioStrain Chip" that employs nanotechnology to induce genetic changes inside the carrier's body.

This miracle chip is advertised as a cure for the H5N1/ebola chimera that is released in the prologue to the book. Of course, if you've read the novel, then you know the Bio-Strain Chip does far more than "cure"—it also kills.

Though a work of fiction, *The Armageddon Strain* raises a chilling question: What limitations lie within the payload of a biochip? Can such a tiny device do more than carry digitized information? Could it actually serve as the *Mark of the Beast*?

The answer is yes.

DNA (Deoxyribonucleic acid) has become the darling of researchers who specialize in synthetic constructs. The "sticky-end" design of the DNA double-helix makes it ideal for use in computing. Though an infinite number of polyhedra are possible, the most robust and stable of these "building blocks" is called the double crossover (DX). An intriguing name, is it not? The double-cross.

Picture an injectible chip comprised of DNA-DX, containing instructions for a super-soldier. Picture, too, how this DNA framework, if transcribed, might also serve a second, *sinister*, purpose—not only to instruct, but also to *alter*.

Mankind has come perilously far in his search for perfection through chemistry. Although millennia passed with little progress beyond roots, herbs, and alchemical quests for gold from lead, the twentieth century ushered science into the rosy dawn of breathless discovery. Electricity, lighter-than-air travel, wireless communication, and computing transformed the ponderous pace of the scientific method into a light-speed race toward self-destruction.

By the mid-1950s, Watson and Crick had solved the structure of the DNA molecule and the double helix became all the rage. Early gene splicing, and thus transgenics, began in 1952 as a crude, cut-and-paste sort of science cooked up in kitchen blenders and petri dishes—as much accident as inspiration. As knowledge has increased (Daniel 12:4), genetic scientists learned to utilize microbiological "vectors" and sophisticated methods to insert animal or plant genes from one species into another. It's the ultimate "Mr. Potato Head" game, where interchangeable plastic pieces give rise to an infinite number of combinations; only, in genetic splicing, humanity is the unhappy potato.

Vectors provide the means of transport and integration for this brave new science. Think of these vectors as biological trucks that carry genetic building materials and workers into your body's cells. Such "trucks" could be a microsyringe, a bacterium, or a virion (a virus particle). Any entity that can carry genetic information (the larger the load capacity, the better) and then surreptitiously gain entry into the cell is a potential vector. Viruses, for example, can be stripped of certain innate genes that might harm the cell. Not only does this (supposedly) render the viral delivery truck "harmless," it also clears out space for the cargo.

Once inside the cell, the "workers" take over. Some of these "workers" are enzymes that cut human genes at specific sites, while others integrate—or load—the "cargo" into appropriate reading frames—like microscopic librarians. Once the payload is stored in the cell's nuclear "library stacks," the new genes can be translated, copied, and "read" to produce altered or brand-new, "alien" polymers and proteins.

The resulting hybrid cell is no longer purely human. If [it is] a hybridized skin cell, it may now glow, or perhaps form scales rather than hair, [or] claws rather than fingernails. If [it is] a brain cell, the new genetic instructions could produce an altered neurotransmitter that reduces or even eliminates the body's need for sleep. Muscle cells may grow larger and more efficient at using low levels of calcium and oxygen. Retina cells may encode for receptors that enable the "posthuman being" to perceive infrared or ultraviolet light frequencies. The hybrid ears may now sense a wider range of sounds, taste buds a greater range of chemicals. Altered brains might even attune to metaphysics and "unseen" gateways, allowing communication with supernatural realms.

Germ-line alterations, mentioned earlier, form a terrifying picture of generational development and may very well already be a reality. Genetic "enhancement" of sperm-producing cells would change human sperm into tiny infiltrators, and any fertilized ovum a living chimera. Science routinely conducts experiments with transgenic mice, rats, chickens, pigs, cows, horses, and many other species. It is naïve to believe humans have been left out of this transgenic equation.

If so many scientists (funded by government entities) believe in the "promise" of genetic alteration and transgenic "enhancement," how then can humanity remain human? We cannot. We will not. Perhaps, *some have not.*

Spiritually, the enemy has ever sought to corrupt God's plan. Originally, fallen angels lay with human women to corrupt the original base pair arrangements. Our genome is filled with "junk DNA" that seemingly encodes for nothing. These "introns" may be the remains of the corrupted genes, and God Himself may have switched them off when fallen angels continued their program, post-Flood. If so, today's scientists might need only to "switch them back on" to resurrect old forms such as Gibborim and Nephilim.

I should point out that not all "trucks" (vectors) deliver their payload immediately. Some operate on a time delay. Cytomegalovirus (CMV) is a common infective agent resident in the cells of many humans today. It "sleeps" in our systems, waiting for a window of opportunity to strike. Recently, genetic specialists began utilizing CMV vectors in transgenic experiments. In 1997, the Fox television program *Millennium* featured an episode in the second season called "Sense and Antisense" (referring to the two sides of the DNA molecule). In this chilling story, a scientist named Lacuna reveals a genetic truth to Frank Black: "They have the map, the map, they can make us go down any street they want to. Streets that we would never even dream of going down. They flip a switch, we go east. They flip another switch, we go north. And we never know we have been flipped, let alone know how."

In the final days of this current age, humanity may indeed "flip." Paul tells us that Christians will be transformed in a moment (1 Corinthians 15:51–53). Is it possible that the enemy also plans an instantaneous "flip"? Are genetic sleeper agents (idling "trucks") already at work in humanity's DNA, waiting and ready to deploy at the appropriate moment?

Science is ready. Knowledge has been increased. The spiritual players have taken the stage.

All we need is the signal. The sign. The injection. The mark. The *trigger*.

We shall ALL be changed. Some to incorruptible bodies ready to meet the Lord. Others to corrupted genomes ready to serve the Beast.

The Man of Sin—and His 666 Mark—Are Coming

Everything we have discussed between the covers of this book foreshadows a very near future in which a man of horrendous, yet unseen, intelligence and diplomacy will emerge on the world scene as a savior. Though his arrival in the form of a man was foretold by numerous Scriptures, the broad masses will not immediately recognize him for what he actually is—paganism's ultimate incarnation, the "Beast" of Revelation 13:1. As he makes himself known, scholars including some of the most celebrated Christian leaders will herald his uncanny ability at resolving whatever emergency gives rise to his appearance, assuring congregations of his godliness and scoffing at those who warn against receiving his "Mark." Only when it is too late will his profound popularity be understood for the ruse it actually is—an unmerciful plot by a very old, superintelligent spirit who ultimately becomes "a king of fierce countenance" (Daniel 8:23). En route to causing all, both small and great, to receive his beastly Mark, the combined depravities of Antiochus Epiphanes, Adolf Hitler, Joseph Stalin, and Genghis Khan, all of whom were types of the Antichrist, will look like child's play in comparison to his brutality. With imperious decree, he will facilitate a One World Government, universal religion, and global socialism. Those who refuse his New World Order will inevitably be imprisoned or destroyed until finally he raises his fist, "speaking great things...in blasphemy against God, to blaspheme his name, and his tabernacle, and them that dwell in heaven" (Revelation 13:5–6). Ultimately, he will exalt himself "above all that is called God, or that is worshipped" until finally he enthrones himself "in the temple of God, showing himself that he is God" (2 Thessalonians 2:4).

Will you be ready to recognize this end-times great deceiver and to escape the mass delusion that will befall "all them that dwell on the face of the earth" (Luke 21:35)?

Will you be "accounted worthy to escape all these things that shall come to pass" (Luke 21:36)?

There is only one way to know for sure. Accept Jesus Christ as your Lord and Savior and repent of your sins. If you will or have done this, He promises: "Because thou hast kept the word of my patience, I also will keep thee from the hour of temptation, which shall come upon all the world, to try them that dwell upon the earth" (Revelation 3:10).

And they overcame him by the blood of the Lamb, and by the word of their testimony; and they loved not their lives unto the death. (Revelation 12:11)

Our genes held together by the grace of God pg. 160

Number 13 pg. 161

Cyborg Transhumanist movies pg 166 252

Voice to skull transmission pg 193

(C) How transhumanists see sin pg 246

Psychotropic drugs pg 250

Nikola Tesla pg 268

Brain activity map chg T.V. by thinking about it pg. 273

Viruses causing mark of the Beast pg 275

Fallen angels with women pg. 281

antichrist pg. 282

You can buy wristband at Amazon pg 209

Avatar "Image to the Beast" pg. 216

Fall of christians pg. 218